10 Texas STAAR Grade 6 Math Practice Tests

The Ultimate Test Prep Collection with Answer Explanations

Dr. A. Nazari

10 Practice Tests

The Grand Championship Collection

Welcome, future Math Champion!

You hold the **ultimate collection** *—*
ten full-length practice tests designed to take you
from first attempt to **complete mastery.**

Conquer every Grade 6 topic

Build unshakeable confidence

Rise from Bronze to Gold to Champion

Arrive at test day fully prepared

The championship begins now.

> **❝** *Ten tests may seem like a marathon, but champions are made one step at a time. Trust the process!* **❞**

# 👑 The Champion's Path 👑

Your 4-phase journey from Bronze to Champion

🥉 Bronze Round (Tests 1–3)

Your warm-up matches. Take these **untimed** to learn the format and set your baseline. Read the answer explanations after each test — this is where you build your foundation.

🥈 Silver Round (Tests 4–6)

Set a timer for **75 minutes**. Focus on the topics that tripped you up in Bronze. Practice showing your work on every problem. Your accuracy should be climbing.

🥇 Gold Round (Tests 7–9)

Full timed conditions (**60 minutes**). Simulate the real exam environment. Review only the questions you missed — targeted practice is the key to gold.

👑 Championship Final (Test 10)

Your final match. Full exam conditions — timed, quiet, no breaks. This is your victory lap. Show yourself how far you've come!

🏆 Your Championship Kit

- 🥇 **10 Full-Length Practice Tests** — every Grade 6 topic
- 🥇 **Formula Reference Sheet**
- 🥇 **Complete Answer Key** with explanations
- 🥇 **Championship Scoreboard** to track your rise

 **Champion's Tip:** Space your tests 2–3 days apart. Use the days in between for targeted review. By Test 10, you'll be amazed at your transformation.

♛ The Champion's Playbook ♛

Seven rules that separate champions from the rest

I   **Read every question twice.** *The first read tells you the topic. The second tells you exactly what to solve for. Champions never skim.*

II   **Mark the clues.** *Circle key numbers, underline the question, and cross out information that's just there to distract you.*

III   **Choose your strategy.** *Before touching pencil to paper, decide: Am I setting up a ratio? Solving an equation? Finding area? Name the approach.*

IV   **Solve, then match.** *For multiple choice — work the problem on scratch paper first, then find your answer among the choices.*

V   **Eliminate and conquer.** *Cross out obviously wrong answers. If you're left with two, you've already doubled your odds. Make an educated pick.*

VI   **Estimate to verify.** *After solving, ask: "Is this answer reasonable?" A quick mental estimate catches most calculation errors.*

VII   **Leave nothing blank.** *Even a well-reasoned guess is worth more than an empty space. Use partial work to support your answer.*

🕐 Timing Mastery

Tests 1–3: **Untimed** (build foundation)   ›   *Tests 4–6:* **75 min** (build speed)   ›   *Tests 7–10:* **60 min** (championship conditions)

★ Grade 6 Championship Topics

🏅 Ratios & Proportions   🏅 Integers & Rational Numbers   🏅 Expressions & Equations

🏅 Geometry & Measurement   🏅 Statistics & Data Analysis

*A true champion isn't someone who never makes mistakes — it's someone who learns from **every single one**. After each test, review your errors carefully. That's where the real growth happens.*

Find more at
ViewMath.com/TX-Grade6

 The Champion's Toolkit

Prepare your workspace before each championship round

🏆 Required Equipment

Sharpened Pencils — Two #2 pencils — champions always have a backup

Quality Eraser — A clean, soft eraser that won't smudge your work

Scratch Paper — Blank paper for calculations, diagrams, and number lines

Ruler — Essential for geometry and coordinate plane questions

Timer — Begin using from the Silver Round onward

Quiet Workspace — A calm, well-lit area free from distractions

🏅 Permitted in Competition

- ✔ Pencil and eraser
- ✔ Scratch paper (provided)
- ✔ Ruler (if specified)
- ✔ Formula reference in this book

🚫 Not Permitted

- ✘ Calculators
- ✘ Electronic devices
- ✘ Textbooks or notes
- ✘ Outside help

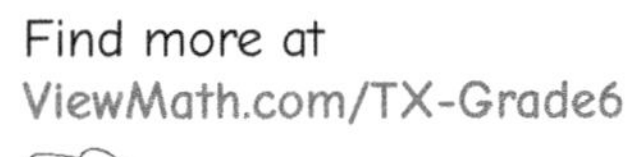

Find more at
ViewMath.com/TX-Grade6

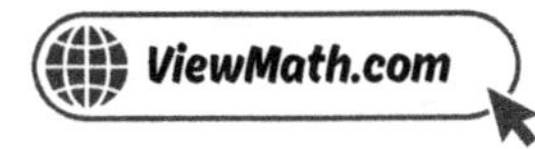

Formula Reference Sheet

◭ Area Formulas

Rectangle	$A = l \times w$
Parallelogram	$A = b \times h$
Triangle	$A = \dfrac{1}{2} \times b \times h$
Trapezoid	$A = \dfrac{1}{2}(b_1 + b_2) \times h$

▣ Volume

Rectangular Prism $V = l \times w \times h$

▣ Surface Area

Find the area of each face, then add them all up.

Rectangular Prism:

$SA = 2lw + 2lh + 2wh$

↓⅑ Order of Operations

P Parentheses first

E Exponents

M/D Multiply & Divide (left to right)

A/S Add & Subtract (left to right)

٪ Ratios & Percents

Ratio: $a : b$ or $\dfrac{a}{b}$

Unit rate: amount per 1 unit

Percent: a ratio out of 100

Part = Percent × Whole

⚖ Integers & Absolute Value

Integers:

$$\ldots, -3, -2, -1, 0, 1, 2, 3, \ldots$$

$|-5| = 5 \quad |5| = 5$

Absolute value = distance from 0

X¹ Expressions & Equations

Exponent: $3^4 = 3 \times 3 \times 3 \times 3 = 81$

Variable: a letter that stands for a number

Equation: two expressions joined by $=$

Inequality: uses $<, >, \leq, \geq$

⊕ Coordinate Plane

Ordered pair: (x, y)

x-axis: horizontal y-axis: vertical

Origin: $(0, 0)$

Four quadrants (I, II, III, IV)

▦ Statistics

Mean: sum of values ÷ count

Median: middle value (sorted)

Range: max − min

Championship Scoreboard

Track your rise through every championship round

Champion's Name: _______________________________

Round	Tier	Date	Score	Rating
1	Bronze		/	
2	Bronze		/	
3	Bronze		/	
4	Silver		/	
5	Silver		/	
6	Silver		/	
7	Gold		/	
8	Gold		/	
9	Gold		/	
10			/	

My strongest topics (where I consistently score well):

____

____

Topics I improved on the most from Bronze to Gold:

____

____

My score trend (Bronze avg → Gold avg → Championship):

____

____

One strategy that helped me improve the most:

____

____

My confidence level for the real test (1–10): _________ / 10

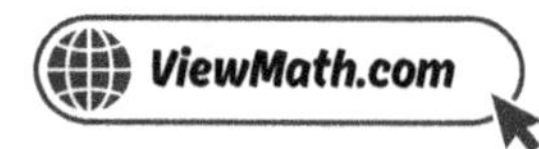

⭐ Table of Contents ⭐

Here's what we'll explore together!

Let's learn and have fun!

Practice Test 1

 30 Questions

✏️ Before You Start ✏️

- ✔ **Read each question carefully** before choosing your answer.
- ✔ **Show your work** on scratch paper when you need to.
- ✔ **Skip hard questions** and come back to them later.
- ✔ **Check your answers** when you're done.
- ✔ **Take your time** — there's no rush!

⭐ You've Got This! ⭐

Do your best and show what you know!

1. A garden has flowers and vegetables in a ratio of 4 : 1. There are 20 flowers. How many vegetables are there?

Your Answer:

2. The graph below shows the cost of buying notebooks at a school supply store.

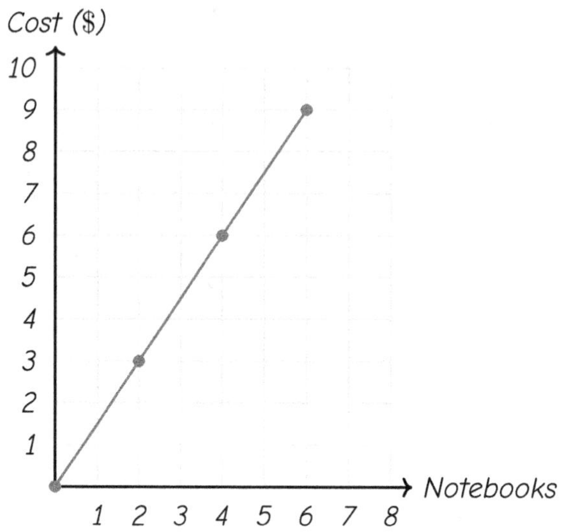

What is the unit price per notebook?

(A) $1.00

(B) $1.25

(C) $1.50

(D) $2.00

3. The double number line below shows equivalent ratios of scoops of mix to cups of water.

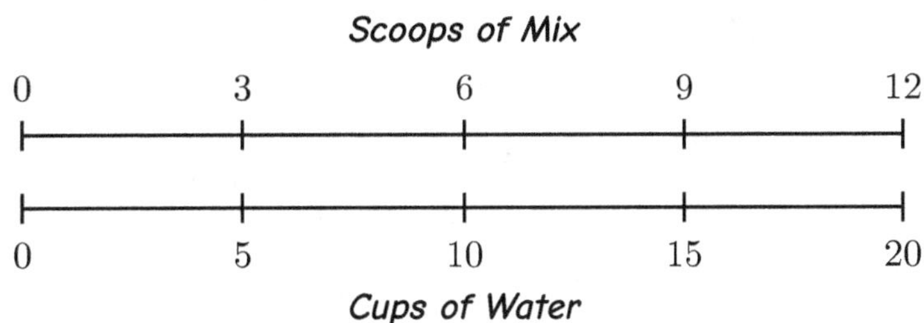

How many cups of water go with 9 scoops of mix?

(A) 10

(B) 12

(C) 15

(D) 18

4. Write 0.15 as a percent.

(A) 1.5%

(B) 15%

(C) 150%

(D) 0.15%

5. The bar model below compares different length units.

1 yard		
1 ft	1 ft	1 ft

If a rope is 7 yards long, how many feet is it?

(A) 10

(B) 14

(C) 21

(D) 28

6. Lily receives \$60 for her birthday. She puts 40% in savings, spends 25% on a book, and donates 10% to charity. How much money does Lily have left?

Your Answer:

7. The table and graph below represent a relationship between x and y.

x	y
0	0
1	3
2	6
3	9
4	?

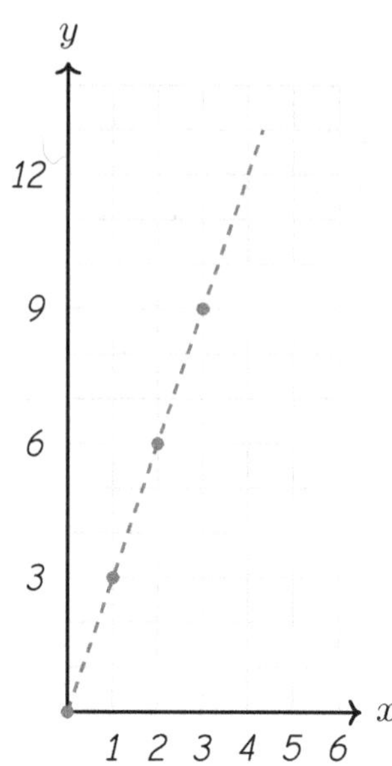

What is the missing value of y when $x = 4$, and is this relationship proportional?

- (A) $y = 10$; non-proportional
- (B) $y = 12$; proportional
- (C) $y = 12$; non-proportional
- (D) $y = 13$; proportional

8. You deposit \$500 in a savings account that earns 4% simple interest per year. How much interest do you earn after 2 years?

- (A) \$20
- (B) \$40
- (C) \$50
- (D) \$80

9. Use the rules for adding integers to find the sum: $(-9) + 4$.

Your Answer:

10. Find $|-32|$.

Your Answer:

11. The number line below shows the subtraction $5 - 9$.

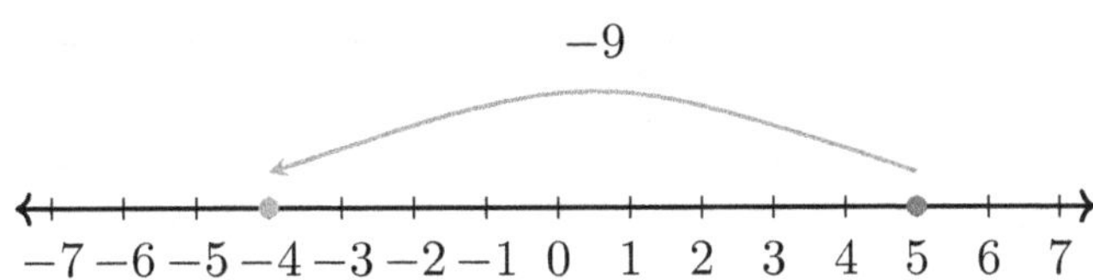

What is $5 - 9$?

(A) 4

(B) −4

(C) −14

(D) 14

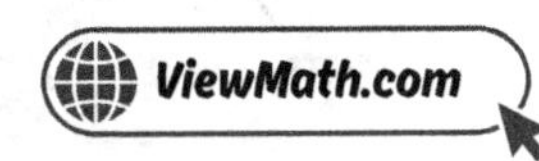

12. A weather station recorded the hourly temperature change in a city on a winter night. The table below shows the temperature change each hour.

Hour	Change per Hour (°F)	Total Change (°F)
1	-3	-3
2	-3	-6
3	-3	-9
4	-3	?
5	-3	?
6	-3	?

Part A: Complete the table for hours 4, 5, and 6.

Part B: Write a multiplication expression that gives the total change after 6 hours, and find the result.

Part C: If the temperature was $20\,°F$ at the start, what is the temperature after 6 hours?

Your Answer:

13. Evaluate: $5^2 + 4^2$

Your Answer:

14. Which expression represents "a number n plus 8"?

(A) $n - 8$

(B) $8n$

(C) $n + 8$

(D) $n \div 8$

15. In the expression $12k + 6$, what are the factors of the term $12k$?

Your Answer:

Find more at
ViewMath.com/TX-Grade6

ViewMath.com

16. You have $50 and spend d dollars. Which expression shows how much money you have left?

(A) $50 + d$

(B) $d - 50$

(C) $50d$

(D) $50 - d$

17. Write and solve an equation: A number decreased by 19 is 31.

Your Answer:

18. Is $n = 4$ a solution to $n > 4$? Is $n = 4$ a solution to $n \geq 4$? Explain both.

Your Answer:

19. Which value can you use to test that the graph of $x > -5$ is correct?

(A) $x = -5$ (it should make the inequality true)

(B) $x = -6$ (it should make the inequality true)

(C) $x = 0$ (it should make the inequality true)

(D) $x = -10$ (it should make the inequality true)

20. A triangular sail has a base of 3 m and a height of 7 m. How much fabric is needed to make 2 identical sails?

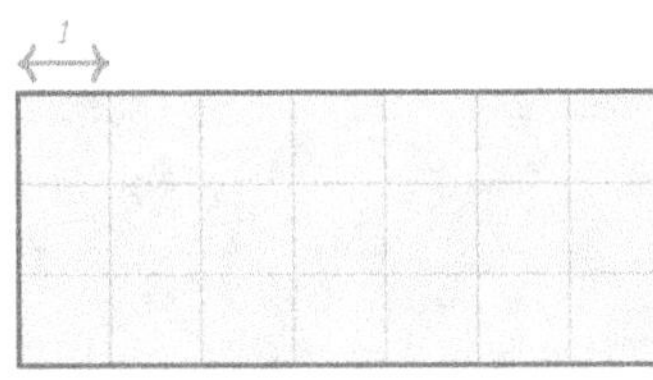

(A) $10.5 \ m^2$

(B) $42 \ m^2$

(C) $21 \ m^2$

(D) $20 \ m^2$

21. Volume is measured in which type of units?

   (A) Square units (cm^2)

   (B) Linear units (cm)

   (C) Cubic units (cm^3)

   (D) No units are needed

22. Points $A(0, 4)$ and $B(0, -5)$ are connected by a line segment. What is the length of $\overline{AB}$?

   (A) 1 unit

   (B) 5 units

   (C) 4 units

   (D) 9 units

23. A right triangle has vertices $(2, 1)$, $(2, 7)$, and $(8, 1)$. What is the area?

   (A) 36 square units

   (B) 18 square units

   (C) 12 square units

   (D) 24 square units

24. A cube has edge length 3 cm. What is the surface area?

   (A) 9 cm^2

   (B) 27 cm^2

   (C) 36 cm^2

   (D) 54 cm^2

25. Data: $10, 20, 30, 40, 50, 60$. What is the median?

   (A) 30

   (B) 33

   (C) 35

   (D) 40

Find more at
ViewMath.com/TX-Grade6

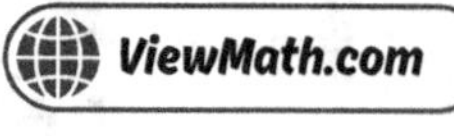

26. The table below shows the distances (in miles) five students travel to school and their distances from the mean.

Student	Miles	Distance from Mean
A	2	4
B	4	2
C	6	0
D	8	2
E	10	4

What is the MAD?

(A) 0

(B) 2

(C) 2.4

(D) 6

27. A histogram has bars for 0–9, 10–19, 20–29, 30–39 with heights $3, 10, 8, 4$. Which interval contains the most data?

(A) 0–9

(B) 10–19

(C) 20–29

(D) 30–39

28. A box plot shows: $min = 20$, $Q_1 = 35$, $median = 50$, $Q_3 = 60$, $max = 95$. Find the range and IQR.

Your Answer

29. School A has 150 sixth graders with a median reading score of 290 and IQR of 30. School B has 100 sixth graders with a median reading score of 285 and IQR of 45. Which school's students are more spread out in reading ability?

(A) School A

(B) School B

(C) They are equally spread.

(D) Cannot be determined without the means.

Find more at
ViewMath.com/TX-Grade6

30. *What is the mode of the data in the stem-and-leaf plot below?*

Stem	Leaves
3	2 5 5 8
4	1 5 5 5 9
5	0 3

Key: 3 | 2 *means* 32

Your Answer:

Find more at
ViewMath.com/TX-Grade6

 # ★ *End of Practice Test 1* ★ 

Great job finishing the test!

🗒️ *My Score*

I got _____________ out of 30 questions right.

*Check your answers in the **Answer Key** at the back of the book.*

💡 Review any questions you missed. That's how we learn!

📊 *Check Your Score Online!*

*Visit **ViewMath Academy** to enter your answers and see which topics you need to review. You can also explore lessons, take quizzes, track your scores, and save your progress!*

viewmath.com/score/6.1.TX.16

Or go to viewmath.com/score and enter code: 6.1.TX.16

Practice Test 2

 30 Questions

 Before You Start 

- ✓ **Read each question carefully** before choosing your answer.
- ✓ **Show your work** on scratch paper when you need to.
- ✓ **Skip hard questions** and come back to them later.
- ✓ **Check your answers** when you're done.
- ✓ **Take your time** — there's no rush!

 You've Got This! 

Do your best and show what you know!

1. Look at the tape diagram below.

Boys: ▨▨▨▨
Girls: ▢▢▢▢▢▢

There are 24 girls. How many boys are there?

(A) 4

(B) 12

(C) 16

(D) 20

2. The table below shows prices for two electric companies.

Company	kWh Used	Monthly Cost
SunPower	800	$96
BrightLight	1,200	$132

Part A: Find the cost per kWh for each company.

Part B: If a family uses 1,000 kWh per month, how much would they pay with each company?

Part C: Which company is the better deal?

Your Answer:

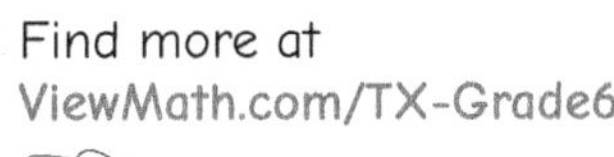

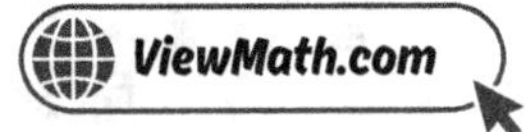

3. The graph below shows points that represent equivalent ratios of flour to sugar in a recipe.

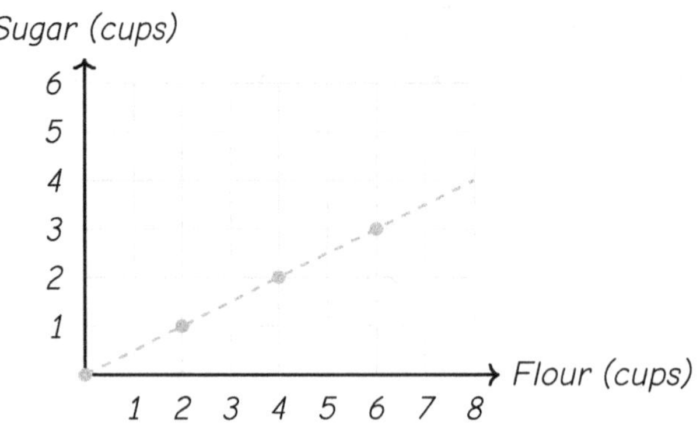

Part A: What is the ratio of flour to sugar?

Part B: If you need 8 cups of flour, how many cups of sugar do you need?

Your Answer:

4. A meal costs $25. You leave a 20% tip. How much is the tip?

A) $3

B) $4

C) $5

D) $6

5. A bag weighs 500 grams. What is its weight in kilograms?

A) 50

B) 5

C) 0.5

D) 0.05

6. Which is the best reason to save money regularly?

A) So you can spend more on wants right away

B) So you have money for unexpected expenses or future goals

C) So the bank will give you a credit card

D) So you never have to earn income again

Find more at
ViewMath.com/TX-Grade6

ViewMath.com

7. Which equation represents a **proportional** relationship?

(A) $y = 4x + 1$

(B) $y = 7x$

(C) $y = x - 3$

(D) $y = 2x + 5$

8. A family earns \$3,200 per month. Their monthly expenses are: rent \$960 (fixed), car payment \$320 (fixed), groceries \$480 (variable), entertainment \$240 (variable), and utilities \$320 (fixed). The rest goes to savings. What is the savings rate as a percent of income?

Your Answer:

9. Which real-world situation is best described by a positive number?

(A) A debt of \$40

(B) 15 feet below sea level

(C) A deposit of \$60 into a savings account

(D) A temperature of 8° below zero

10. If $a + b = 0$ and $a = -12$, what is the value of b?

Your Answer:

11. At noon, the temperature was 5°C. By midnight, it had dropped 12°C. What was the temperature at midnight?

(A) 17°C

(B) 7°C

(C) −7°C

(D) −17°C

12. What is $(-120) \div 6$?

Your Answer:

13. *A student is building a cube out of unit cubes. Each edge of the cube is 3 units long. The diagram shows the base layer.*

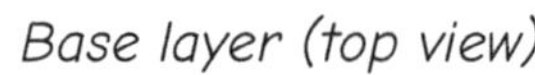
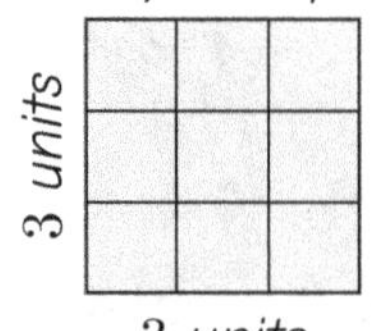

Base layer (top view)

3 units

3 units

How many unit cubes are needed to build the full cube? Write your answer using an exponent.

Your Answer

14. *You earn $12 per hour and worked h hours. You also got a $20 bonus. Write an expression for your total pay.*

Your Answer

15. *What is the coefficient of y in the expression $8 + 3y$?*

(A) 3

(B) 8

(C) y

(D) 11

16. *In the formula $P = 4s$, the variable s represents the side length of a square. What does P represent?*

(A) *The area of the square*

(B) *The perimeter of the square*

(C) *The number of sides*

(D) *The diagonal of the square*

17. *Solve: $k + 15 = 32$*

(A) $k = 47$

(B) $k = 17$

(C) $k = 27$

(D) $k = 2$

Find more at
ViewMath.com/TX-Grade6

ViewMath.com

18. Which of the following is a solution to $n \leq 8$?

(A) $n = 9$

(B) $n = 8.5$

(C) $n = 8$

(D) $n = 10$

19. Write the inequality that matches this description: closed circle at -4, shade to the left.

Your Answer:

20. A triangle has base 14 m and height 8 m. What is its area?

(A) $112\ m^2$

(B) $22\ m^2$

(C) $44\ m^2$

(D) $56\ m^2$

21. A storage container is 5 m long, 3 m wide, and 2 m tall. It is half-full of sand. How many cubic meters of sand are in the container?

(A) $30\ m^3$

(B) $15\ m^3$

(C) $10\ m^3$

(D) $60\ m^3$

22. A rectangle has vertices $(-6, 2)$, $(2, 2)$, $(2, -3)$, $(-6, -3)$. What is the perimeter?

Your Answer:

23. To find the area of an irregular polygon on the coordinate plane, you can:

(A) Multiply all the coordinates together.

(B) Break it into rectangles and triangles, then add the areas.

(C) Count only the vertices.

(D) Subtract the perimeter from the largest coordinate.

24. A rectangular prism has dimensions 2 ft by 3 ft by h ft. Its surface area is 62 ft^2. What is h?

Your Answer:

25. The bar graph below shows test scores for 5 students.

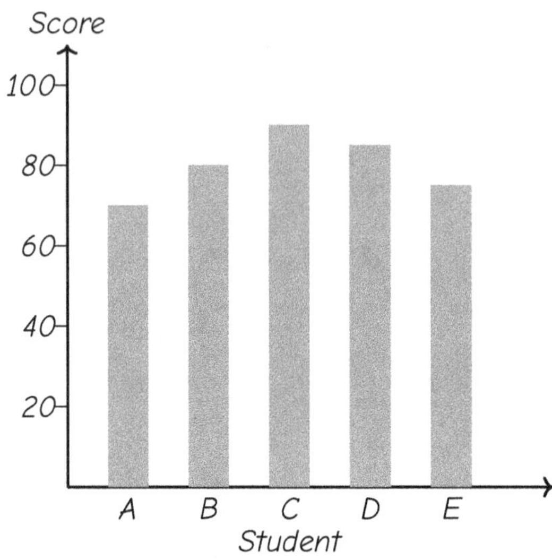

The scores are 70, 80, 90, 85, 75. What is the mean score?

(A) 75

(B) 80

(C) 85

(D) 90

26. The dot plot below shows the number of laps students ran during gym class.

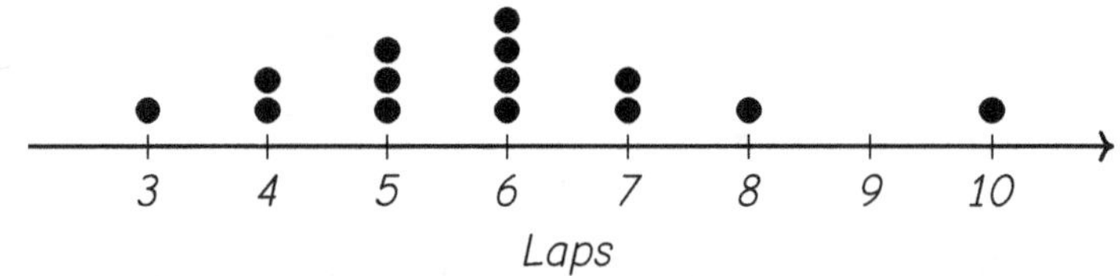

Find the range and IQR from the dot plot data.

Your Answer:

27. A frequency table lists values and how many times each appears. Another name for the value that appears most often is —

(A) mean

(B) median

(C) mode

(D) range

28. Which part of a box plot can you NOT determine from a five-number summary alone?

(A) The median

(B) The IQR

(C) The number of data points

(D) The range

29. A dot plot shows test scores for Class 1: most dots clustered between 75 and 85 with one dot at 40. Class 2's dot plot shows dots evenly spread from 55 to 95. Which class is likely more consistent?

(A) Class 1, because most scores are close to-gether.

(B) Class 2, because the scores are evenly spread.

(C) Neither — they are equally consistent.

(D) Cannot be determined without the means.

30. The back-to-back stem-and-leaf plot below shows the test scores for two classes.

Class A	Stem	Class B
8 5 2	6	3 6
7 4 1	7	0 2 5 9
6 0	8	1 4
	9	2

Key: 2 | 6 | 3 means 62 for Class A and 63 for Class B

Which class has the higher median score?

(A) Class A

(B) Class B

(C) They have the same median.

(D) There is not enough information to tell.

Find more at
ViewMath.com/TX-Grade6

 End of Practice Test 2 

Great job finishing the test!

 My Score

I got __________ out of 30 questions right.

Check your answers in the **Answer Key** at the back of the book.

Review any questions you missed. That's how we learn!

📊 Check Your Score Online!

Visit **ViewMath Academy** to enter your answers and see which topics you need to review. You can also explore lessons, take quizzes, track your scores, and save your progress!

viewmath.com/score/6.1.TX.17

Or go to viewmath.com/score and enter code: 6.1.TX.17

Practice Test 3

30 Questions

✏️ Before You Start ✏️

- ✔ **Read each question carefully** before choosing your answer.
- ✔ **Show your work** on scratch paper when you need to.
- ✔ **Skip hard questions** and come back to them later.
- ✔ **Check your answers** when you're done.
- ✔ **Take your time** — there's no rush!

⭐ You've Got This! ⭐

Do your best and show what you know!

1. A baker uses 2 eggs **per** 3 cups of flour. Which ratio represents flour to eggs?

   (A) $2:3$ (B) $3:5$

   (C) $3:2$ (D) $2:5$

2. Brand A sells 6 granola bars for $5.10. Brand B sells 8 granola bars for $6.00. Which is the better deal?

   (A) Brand A ($0.85 per bar)   (B) Brand B ($0.75 per bar)

   (C) They cost the same   (D) Brand A ($0.75 per bar)

3. A store sells 7 bananas for $2. How much do 21 bananas cost?

   (A) $4   (B) $6

   (C) $9   (D) $14

4. A pair of shoes costs $80. They are 15% off. What is the sale price?

   (A) $60   (B) $65

   (C) $68   (D) $72

5. Convert 156 inches to feet.

   (A) 11   (B) 12

   (C) 13   (D) 14

6. Mia earns $120 babysitting. She wants to save 25% of her earnings. How much should she save?

   (A) $20   (B) $25

   (C) $30   (D) $35

7. The ordered pair $(4, 10)$ belongs to a proportional relationship. What is the constant of proportionality?

(A) $\dfrac{2}{5}$

(B) 2.5

(C) 4

(D) 14

8. Maria earns $600 per month. She spends $450 on expenses. How much can she save?

(A) $50

(B) $100

(C) $150

(D) $200

9. Mia has $15 in her bank account and writes a check for $20. Which expression and result show her new balance?

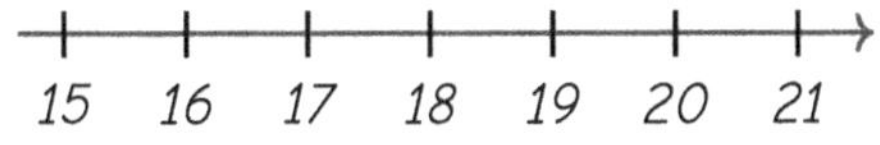

(A) $15 + 20 = 35$

(B) $15 + (-20) = -5$

(C) $(-15) + 20 = 5$

(D) $(-15) + (-20) = -35$

10. A charity tracked its monthly balance: $-\$40$, $\$30$, $-\$55$, $\$10$. Which month had the largest distance from $0?

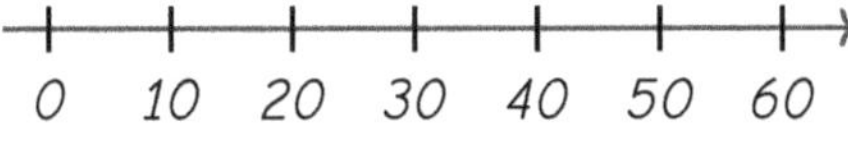

(A) Month 1 $(-\$40)$

(B) Month 2 $(\$30)$

(C) Month 3 $(-\$55)$

(D) Month 4 $(\$10)$

11. The temperature at sunrise was $-8°F$. By noon, it had risen $15°F$. What was the temperature at noon?

(A) $-23°F$

(B) $-7°F$

(C) $7°F$

(D) $23°F$

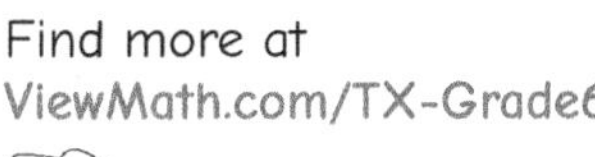
Find more at
ViewMath.com/TX-Grade6

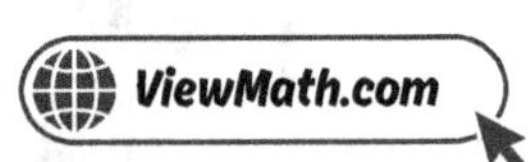

12. A scuba diver descends 4 feet every second. What is the diver's change in elevation after 12 seconds?

Your Answer:

13. What is the value of 10^0?

(A) 0

(B) 1

(C) 10

(D) 100

14. Write an expression for: "the product of 7 and w, increased by 5."

Your Answer:

15. In the term $\dfrac{n}{5}$, what are the factors?

(A) n and 5

(B) n and $\dfrac{1}{5}$

(C) 1 and n

(D) 5 and $\dfrac{1}{n}$

16. A plumber charges \$50 for a house visit plus \$40 per hour of work. Which of the following represents the cost of a 3-hour job?

(A) \$90

(B) \$120

(C) \$170

(D) \$190

Find more at
ViewMath.com/TX-Grade6

ViewMath.com

17. The table shows Mario's steps for solving an equation. In which step did he make an error?

$$\text{Equation:} \quad \frac{n}{5} = 8$$

Step 1: Multiply both sides by 5

Step 2: $n = 8 + 5$

Step 3: $n = 13$

(A) No error — $n = 13$ is correct

(B) Step 1 — he should have divided both sides by 5

(C) Step 2 — multiplying by 5 gives $n = 8 \times 5 = 40$, not $8 + 5$

(D) Step 3 — $8 + 5 = 14$, not 13

18. A parking lot allows no more than 200 cars. Which inequality describes the number of cars c?

(A) $c > 200$

(B) $c \geq 200$

(C) $c < 200$

(D) $c \leq 200$

19. Which of the following graphs represents $x \leq -1$?

(A) Open circle at -1, shade right

(B) Closed circle at -1, shade right

(C) Open circle at -1, shade left

(D) Closed circle at -1, shade left

20. A triangular pennant has a base of 5 in and a height of 12 in. What is its area?

(A) $60 \; in^2$

(B) $17 \; in^2$

(C) $30 \; in^2$

(D) $34 \; in^2$

Find more at
ViewMath.com/TX-Grade6

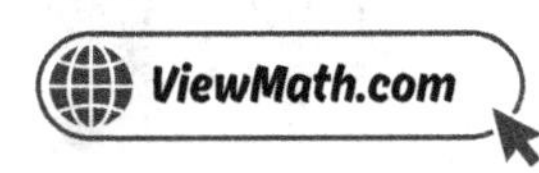

21. A rectangular prism has length $\frac{3}{4}$ ft, width 2 ft, and height 4 ft. What is the volume?

   (A) 6 ft^3

   (B) 3 ft^3

   (C) 24 ft^3

   (D) $\frac{3}{2}$ ft^3

22. Find the perimeter of the rectangle shown on the coordinate plane.

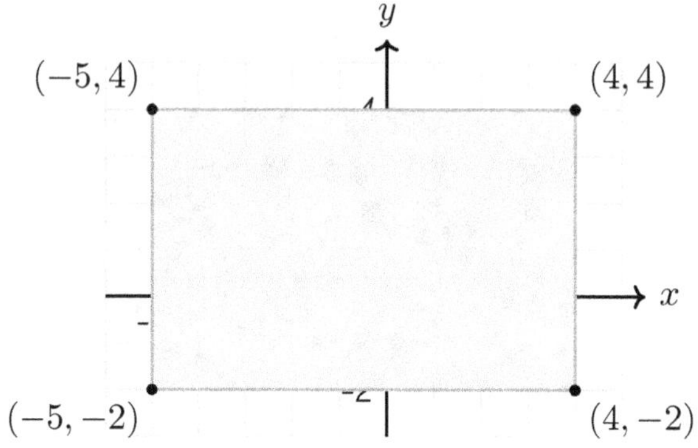

Your Answer

23. A right triangle has vertices $(1, 1)$, $(1, 9)$, and $(5, 1)$. What is the area?

   (A) 32 square units

   (B) 16 square units

   (C) 20 square units

   (D) 8 square units

24. Mia paints all sides of a wooden block that is 3 ft by 2 ft by 4 ft. She then paints an identical block. How much total area does she paint?

   (A) 52 ft^2

   (B) 104 ft^2

   (C) 48 ft^2

   (D) 26 ft^2

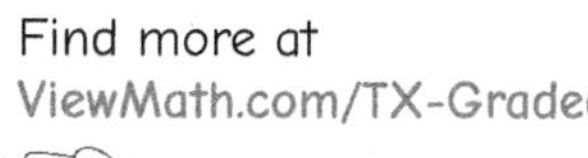

Find more at
ViewMath.com/TX-Grade6

ViewMath.com

25. The mean of 6 quiz scores is 15. If one score of 9 is removed, what is the mean of the remaining 5 scores?

(A) 14.2

(B) 15

(C) 16.2

(D) 18

26. Data: $50, 55, 60, 65, 70$. The mean is 60. Find the MAD.

Your Answer

27. A data set is: $5, 5, 5, 5, 5$. What is the mode?

(A) 0

(B) 1

(C) 5

(D) There is no mode.

28. Data: $10, 12, 14, 15, 18, 20, 22, 25, 30$. What is Q_1?

(A) 12

(B) 13

(C) 14

(D) 15

29. Heights (inches) of Team 1: $60, 62, 63, 64, 65, 66, 68$. Heights of Team 2: $58, 60, 61, 64, 67, 69, 72$. Which team has the greater range?

(A) Team 1

(B) Team 2

(C) They are equal.

(D) Cannot be determined.

30. *A stem-and-leaf plot uses stems* $10, 11, 12$ *with a key* $10 \mid 2 = 102.$ *What data value does stem* 11 *with leaf* 5 *represent?*

(A) 15

(B) 115

(C) 150

(D) 1105

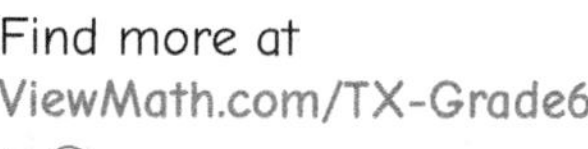

 # End of Practice Test 3 

Great job finishing the test!

✅ My Score

I got _____________ out of 30 questions right.

*Check your answers in the **Answer Key** at the back of the book.*

💡 *Review any questions you missed. That's how we learn!*

📊 Check Your Score Online!

Visit **ViewMath Academy** to enter your answers and see which topics you need to review. You can also explore lessons, take quizzes, track your scores, and save your progress!

viewmath.com/score/6.1.TX.18

Or go to viewmath.com/score and enter code: 6.1.TX.18

Practice Test 4

 30 Questions

✏ Before You Start ✏

- ✔ **Read each question carefully** before choosing your answer.
- ✔ **Show your work** on scratch paper when you need to.
- ✔ **Skip hard questions** and come back to them later.
- ✔ **Check your answers** when you're done.
- ✔ **Take your time** — there's no rush!

 ★ You've Got This!  ★

Do your best and show what you know!

1. A store sells 3 pencils **for every** 1 eraser. Which ratio represents pencils to erasers?

(A) $1:3$

(B) $3:1$

(C) $3:4$

(D) $1:4$

2. An electrician charges $300 for 4 hours of work plus a $50 service fee. What is the cost per hour of labor? What is the total cost per hour (including the service fee)?

Your Answer:

3. The ratio table below has an error in one row. Which row has the error?

x	y
4	10
8	20
12	25
16	40

(A) Row 1

(B) Row 2

(C) Row 3

(D) Row 4

4. A store is having a "Buy More, Save More" sale. Spend $50–$99: get 10% off. Spend $100+: get 20% off. You want to buy items totaling $95. Should you add a $10 item to reach $105? Explain.

Your Answer:

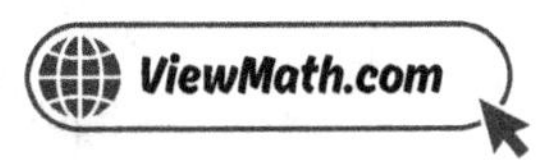

5. How many inches are in 5 feet?

(A) 50

(B) 55

(C) 60

(D) 72

6. When you use a **debit card**, the money comes directly from your —

(A) credit card company

(B) the store's account

(C) a loan from the bank

(D) your own bank account

7. A proportional relationship has $k = \dfrac{3}{4}$. Which ordered pair belongs to this relationship?

(A) $(3, 4)$

(B) $(8, 6)$

(C) $(5, 4)$

(D) $(6, 8)$

8. You deposit $600 into a savings account that earns 5% simple interest per year. How much interest do you earn after 3 years?

Your Answer

9. What is $(-4) + (-3)$?

(A) -7

(B) -1

(C) 7

(D) 1

10. What is $|-7|$?

  (A) -7 (B) 7

  (C) 0 (D) $-(-7)$

11. A hiker stands at an elevation of 120 feet above sea level and descends 185 feet into a canyon. What is the hiker's new elevation?

  (A) 305 feet   (B) 65 feet

  (C) -65 feet   (D) -305 feet

12. What is $(-4) \times (-5) \times 3$?

Your Answer:

13. The table below shows powers of 3. What value belongs in the blank?

3^1	3^2	3^3	3^4	3^5
3	9	27	?	243

  (A) 36   (B) 64

  (C) 81   (D) 108

14. Write an expression for: "a number t divided by 4, then subtract 1."

Your Answer:

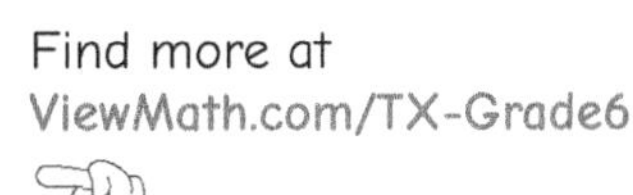

Find more at
ViewMath.com/TX-Grade6

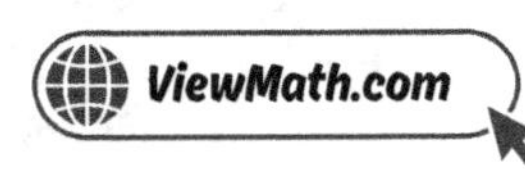

15. How many terms are in the expression $7k$?

    (A) 1   (B) 2

    (C) 3   (D) 7

16. The diagram shows a garden with a fixed fence on the left and new fencing needed for the other three sides. The garden is w feet wide and 12 feet long. Which expression gives the length of **new** fencing needed?

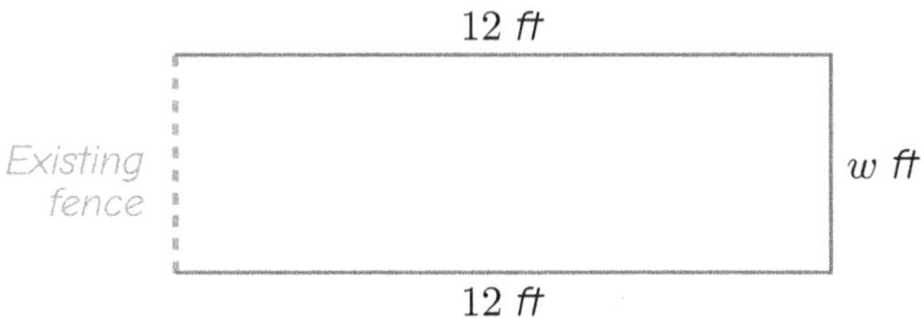

    (A) $24 + w$ (B) $24 + 2w$

    (C) $12 + 2w$ (D) $2(12 + w)$

17. Solve: $12p = 84$

    (A) $p = 72$ (B) $p = 96$

    (C) $p = 7$ (D) $p = 6$

18. Which inequality represents "you need more than \$25 to buy the video game"?

    (A) $d \leq 25$ (B) $d < 25$

    (C) $d > 25$ (D) $d \geq 25$

19. Which inequality has a graph where $x = -4$ is in the shaded region but $x = -2$ is NOT?

(A) $x > -3$

(B) $x < -3$

(C) $x \geq -3$

(D) $x \leq -3$

20. A triangle has base 20 in and height 7 in. What is the area?

(A) $140\ in^2$

(B) $27\ in^2$

(C) $70\ in^2$

(D) $54\ in^2$

21. A rectangular prism has a volume of $120\ cm^3$. Its length is 10 cm and width is 4 cm. What is the height?

(A) 3 cm

(B) 12 cm

(C) 6 cm

(D) 30 cm

22. A rectangle has vertices $(-5, 4)$, $(3, 4)$, $(3, -2)$, and $(-5, -2)$. What is the perimeter?

(A) 14 units

(B) 22 units

(C) 28 units

(D) 48 units

23. What is the area of the shaded right triangle on the coordinate plane?

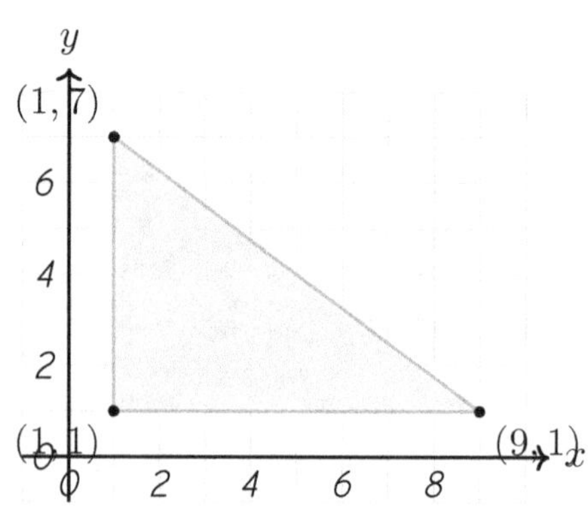

(A) 48 square units

(B) 24 square units

(C) 36 square units

(D) 12 square units

24. A cube has a surface area of 216 cm^2. What is the edge length?

Your Answer:

25. Find the median of $12, 18, 20, 24$

(A) 18

(B) 19

(C) 20

(D) 21

26. Data A: $IQR = 4$. Data B: $IQR = 20$. What can you conclude?

(A) Data A has a higher median.

(B) Data B has a higher median.

(C) The middle 50% of Data A is more tightly clustered than Data B.

(D) Data A has more data points.

27. You want to know whether to display data as a dot plot, histogram, or frequency table. Which question should you ask first?

(A) What is the mean of the data?

(B) How many data points are there, and is the data numerical or categorical?

(C) What color should the bars be?

(D) Is the data symmetric or skewed?

28. A box plot has its median line closer to Q_1 than to Q_3. Which best describes the shape of the data?

(A) Symmetric

(B) Skewed left

(C) Skewed right

(D) Uniform

29. Two box plots are shown for heights (in cm) of plants grown with Fertilizer A and Fertilizer B.

Write the five-number summary for each fertilizer. Compare the medians, ranges, and IQRs. Which fertilizer produced taller plants on average? Which produced more consistent growth?

Your Answer

30. The data set is: $32, 35, 41, 44, 48, 53, 57$. Which stem in the stem-and-leaf plot would have the most leaves?

(A) 3

(B) 4

(C) 5

(D) All stems have the same number of leaves.

 # End of Practice Test 4 

Great job finishing the test!

☑ My Score

I got _____________ out of 30 questions right.

*Check your answers in the **Answer Key** at the back of the book.*

💡 *Review any questions you missed. That's how we learn!*

📊 Check Your Score Online!

Visit **ViewMath Academy** to enter your answers and see which topics you need to review. You can also explore lessons, take quizzes, track your scores, and save your progress!

viewmath.com/score/6.1.TX.19

Or go to viewmath.com/score and enter code: 6.1.TX.19

5

Practice Test 5

 30 Questions

✏️ Before You Start ✏️

✓ **Read each question carefully** before choosing your answer.

✓ **Show your work** on scratch paper when you need to.

✓ **Skip hard questions** and come back to them later.

✓ **Check your answers** when you're done.

✓ **Take your time** — there's no rush!

⭐ **You've Got This!** ⭐

Do your best and show what you know!

1. *A class votes on two activities: hiking and swimming. The ratio of votes for hiking to swimming is 3 : 2. There are 25 votes total. How many voted for swimming?*

Your Answer:

2. *Streaming Service A costs $108 per year. Streaming Service B costs $10 per month. Which service is cheaper per month?*

(A) *Service A*

(B) *Service B*

(C) *They cost the same*

(D) *Not enough information*

3. *It takes 6 cups of flour to make 4 loaves of bread. How many cups of flour for 10 loaves?*

(A) 12

(B) 10

(C) 15

(D) 20

4. *The bar chart shows the original price and sale price of three items.*

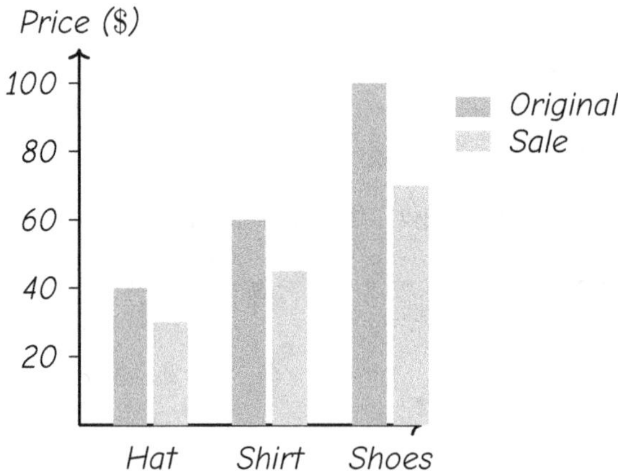

The hat was $40 and is now $30. What is the percent discount on the hat?

(A) 10%

(B) 20%

(C) 25%

(D) 30%

Find more at
ViewMath.com/TX-Grade6

ViewMath.com

5. How many meters are in 4.5 kilometers?

(A) 45

(B) 450

(C) 4,500

(D) 45,000

6. Jordan has a credit card balance of $250. The card charges 15% annual interest. If Jordan pays nothing for 2 years, how much interest will he owe?

Your Answer:

7. The graph of a proportional relationship is always a straight line that passes through which point?

(A) $(1, 1)$

(B) $(0, 1)$

(C) $(1, 0)$

(D) $(0, 0)$

8. Mia earns $1,000 per month. Her budget: rent 35%, utilities 10%, food 20%, transportation 15%, and the rest goes to savings. She gets a raise to $1,200 per month but keeps the same percentages. How much **more** does she save per month after the raise?

(A) $20

(B) $30

(C) $40

(D) $50

9. What is $(-8) + 5$?

(A) -13

(B) 13

(C) -3

(D) 3

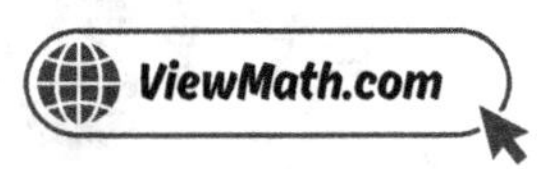

10. Alex has $-\$14$ in his checking account and $-\$9$ in his savings account. Find $|-14| + |-9|$ and explain what the result means.

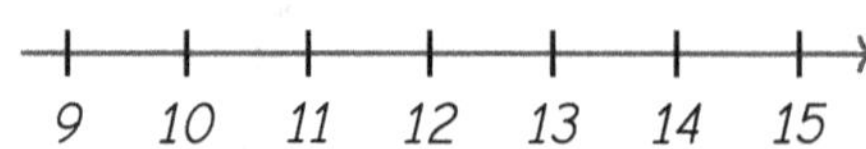

Your Answer

11. Compute $(-14) + (-23)$. Show your work using the rules for adding integers with the same sign.

Your Answer

12. What is $9 \times (-6)$?

(A) 54

(B) -54

(C) -15

(D) 15

13. Evaluate: $(12 - 4)^2 \div 16$

Your Answer

14. Sam has x stickers. He gives away 6. Which expression shows how many stickers Sam has now?

(A) $x + 6$

(B) $6x$

(C) $6 - x$

(D) $x - 6$

15. Which of the following is a factor of the term $9x$?

   (A) $9 + x$ (B) $9x$

   (C) x (D) $9 - x$

16. A phone battery starts at 100% and loses 8% each hour. Write an expression for the battery level after h hours.

Your Answer:

17. Which inverse operation would you use to solve $\dfrac{n}{3} = 12$?

   (A) Divide both sides by 3   (B) Subtract 3 from both sides

   (C) Add 3 to both sides   (D) Multiply both sides by 3

18. Which phrase means $n \geq 12$?

   (A) n is less than 12   (B) n is greater than 12

   (C) n is at least 12   (D) n is at most 12

19. Which of these numbers is a solution to $x \geq -3$?

   (A) -4 (B) -3.5

   (C) -3 (D) -100

20. Two triangles both have a base of 10 cm. Triangle P has a height of 6 cm and Triangle Q has a height of 8 cm. How much greater is the area of Triangle Q?

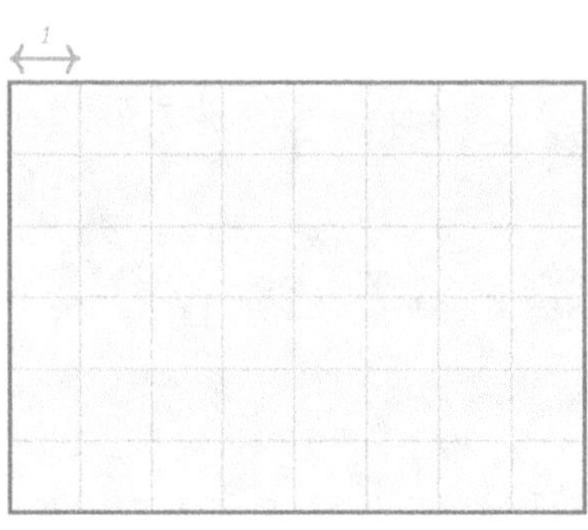

(A) $2\ cm^2$

(B) $10\ cm^2$

(C) $20\ cm^2$

(D) $40\ cm^2$

21. A box is 2.5 cm long, 4 cm wide, and 6 cm tall. What is the volume?

Your Answer:

22. A rectangle has vertices $(1, 2)$, $(7, 2)$, $(7, -4)$, and $(1, -4)$. What is the length of the longer side?

(A) 5 units

(B) 6 units

(C) 7 units

(D) 8 units

23. A rectangle on the coordinate plane has area 54 square units. Two of its vertices are at $(-3, 2)$ and $(6, 2)$. What is the width of the rectangle?

(A) 6 units

(B) 9 units

(C) 3 units

(D) 18 units

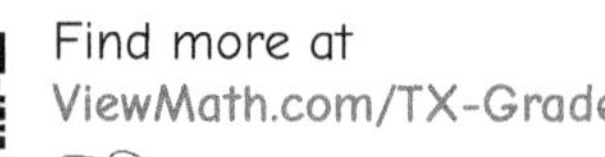

Find more at
ViewMath.com/TX-Grade6

ViewMath.com

24. What is the surface area of a rectangular prism with length 6 cm, width 4 cm, and height 3 cm?

(A) $72 \ cm^2$

(B) $108 \ cm^2$

(C) $54 \ cm^2$

(D) $96 \ cm^2$

25. Which measure of center is more affected by an outlier?

(A) Median

(B) Mean

(C) Both are equally affected

(D) Neither is affected

26. Data: $20, 22, 24, 26, 28$. The mean is 24. What is the MAD?

(A) 2

(B) 2.4

(C) 4

(D) 8

27. A frequency table shows students' favorite lunch: Pizza (12), Chicken (8), Salad (4), Burgers (6). Find the total and the mode.

Your Answer:

28. A student says: "Since the right whisker of this box plot is longer than the left whisker, the right side has more data points." Is the student correct?

(A) Yes — longer whiskers always mean more data.

(B) No — each whisker represents about 25% of the data regardless of length.

(C) Yes — longer whiskers hold twice as much data.

(D) No — whiskers never represent any actual data.

Find more at
ViewMath.com/TX-Grade6

29. *Weekly allowances of two groups of students are listed below.*

Group X: $5, $8, $10, $10, $12, $15, $20

Group Y: $9, $10, $10, $11, $11, $12, $12

Find the median, range, and IQR for each group. Write 2–3 sentences comparing the groups.

Your Answer:

30. *A student made the stem-and-leaf plot below. What is wrong with it?*

Stem	Leaves
5	3 1 7 9
6	2 4

Key: 5 | 3 means 53

(A) *The stems are out of order.*

(B) *The leaves for stem 5 are not in order from least to greatest.*

(C) *The key is incorrect.*

(D) *There are too few stems.*

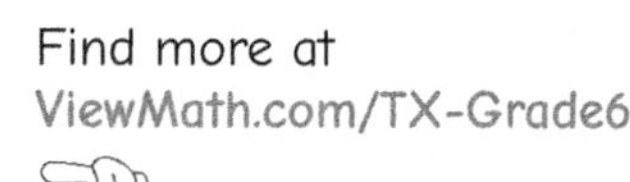

Find more at
ViewMath.com/TX-Grade6

ViewMath.com

 End of Practice Test 5 

Great job finishing the test!

📋 My Score

I got _____________ out of 30 questions right.

*Check your answers in the **Answer Key** at the back of the book.*

💡 *Review any questions you missed. That's how we learn!*

📊 Check Your Score Online!

Visit **ViewMath Academy** to enter your answers and see which topics you need to review. You can also explore lessons, take quizzes, track your scores, and save your progress!

viewmath.com/score/6.1.TX.20

Or go to viewmath.com/score and enter code: 6.1.TX.20

Practice Test 6

 *30 Questions*

 **Before You Start** 

- ✓ **Read each question carefully** before choosing your answer.
- ✓ **Show your work** on scratch paper when you need to.
- ✓ **Skip hard questions** and come back to them later.
- ✓ **Check your answers** when you're done.
- ✓ **Take your time** — there's no rush!

 **You've Got This!** 

Do your best and show what you know!

1. A smoothie recipe uses 3 bananas **for every** 4 cups of yogurt. Write the ratio of bananas to yogurt. Then write the ratio of yogurt to bananas.

Your Answer:

2. A phone plan charges $45 for 3 months. What is the cost per month?

(A) $12

(B) $15

(C) $18

(D) $20

3. A teacher hands out 3 pencils for every 2 students. There are 16 students. How many pencils does she need?

(A) 18

(B) 24

(C) 20

(D) 32

4. A backpack is $60 with a 25% off coupon. What is the sale price?

(A) $35

(B) $40

(C) $45

(D) $50

5. A race is 10 kilometers. How many centimeters is that?

(A) 100,000

(B) 10,000

(C) 1,000,000

(D) 1,000

6. Jake mows lawns and earns $200 per month. He saves 15% each month. How much does he save in 6 months?

(A) $30

(B) $90

(C) $150

(D) $180

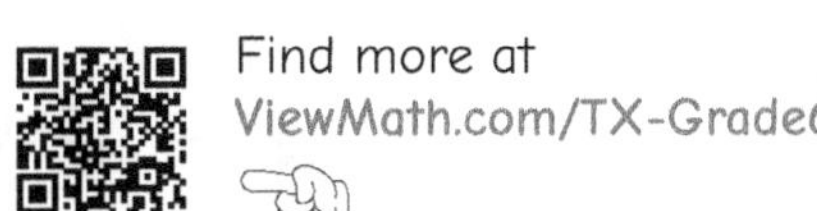
Find more at
ViewMath.com/TX-Grade6

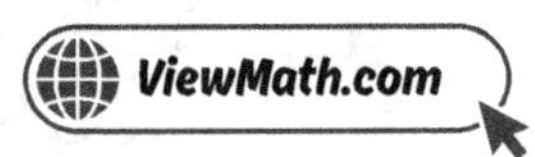

7. A line on a coordinate plane passes through $(0,0)$ and $(1,6)$. Which equation describes this line?

(A) $y = x + 6$

(B) $y = 6x$

(C) $y = x + 5$

(D) $y = 6$

8. A student earns \$120 per week. She puts 25% into savings. After 6 weeks she deposits all her savings into an account that earns 3% simple interest per year. How much interest does she earn in 1 year?

(A) \$3.60

(B) \$5.40

(C) \$9.00

(D) \$18.00

9. The temperature outside is $-15°F$. What does this mean?

(A) 15 degrees above zero

(B) 15 degrees below zero

(C) Exactly zero degrees

(D) 15 degrees above freezing

10. Which number has the greatest absolute value?

(A) 3

(B) -8

(C) -1

(D) 5

11. Which statement about subtracting a negative number is **true**?

(A) $a - (-b)$ is the same as $a - b$

(B) $a - (-b)$ is always negative

(C) $a - (-b)$ is the same as $a + b$

(D) Subtracting a negative gives a smaller result

12. The table below shows the sign rules for multiplying integers. One cell is missing.

Factor 1	Factor 2	Product Sign
+	+	+
+	−	−
−	+	−
−	−	?

What sign belongs in the missing cell?

(A) Negative (−)

(B) Positive (+)

(C) Zero

(D) It depends on the numbers

13. Evaluate: $60 \div (3 + 2) - 2^2$

Your Answer

14. Which expression represents "divide 20 by a number p"?

(A) $20 + p$

(B) $p \div 20$

(C) $20p$

(D) $20 \div p$

15. What is the coefficient of d in the expression $15 - d + 3$?

(A) 0

(B) 1

(C) −1

(D) 15

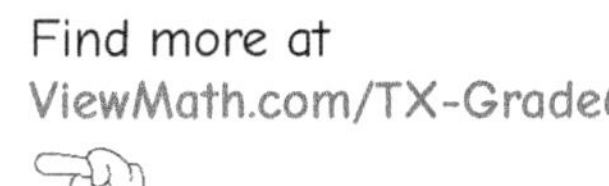
Find more at
ViewMath.com/TX-Grade6

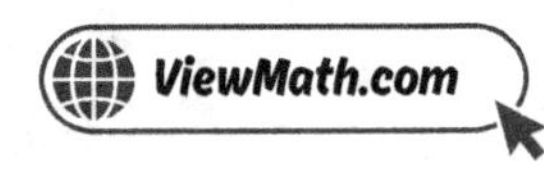

16. A pizza costs $14 and is split equally among f friends. Which expression shows each friend's share?

(A) $14f$

(B) $14 + f$

(C) $14 - f$

(D) $14 \div f$

17. Solve: $6k = 54$. Then check your answer.

Your Answer:

18. Name three solutions to the inequality $m \geq 6$.

Your Answer:

19. What is the difference between the graphs of $x > 5$ and $x \geq 5$?

(A) They shade in different directions

(B) $x > 5$ uses an open circle; $x \geq 5$ uses a closed circle

(C) $x > 5$ shades right; $x \geq 5$ shades left

(D) There is no difference

20. A triangle has base $\frac{3}{4}$ ft and height $\frac{2}{3}$ ft. What is its area?

Your Answer:

21. If you double the height of a rectangular prism but keep the length and width the same, what happens to the volume?

(A) The volume stays the same.

(B) The volume doubles.

(C) The volume triples.

(D) The volume quadruples.

Find more at
ViewMath.com/TX-Grade6

22. Points $A(-1, -4)$ and $B(-1, 8)$ form a vertical segment. What is its length?

Your Answer:

23. A rectangle on the coordinate plane has an area of 56 square units. Its length is 8 units. What is the width?

(A) 6 units

(B) 7 units

(C) 8 units

(D) 48 units

24. How many faces does a rectangular prism have?

(A) 4

(B) 5

(C) 6

(D) 8

25. The mean of five numbers is 20. What is the sum of the five numbers?

(A) 4

(B) 25

(C) 80

(D) 100

26. Team A scores: $60, 62, 64, 66, 68$. Team B scores: $40, 50, 64, 78, 88$. Both have the same mean. Which team is more consistent?

(A) Team A, because its range is smaller.

(B) Team B, because its range is smaller.

(C) They are equally consistent because their means are equal.

(D) Neither is consistent.

Find more at
ViewMath.com/TX-Grade6

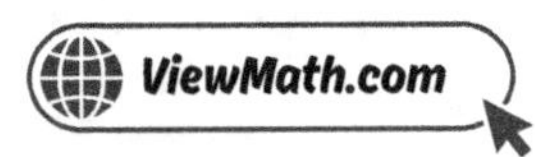

27. In a histogram, there are no gaps between the bars because —

(A) the data values are all the same

(B) the intervals are consecutive and cover every value in the range

(C) the bars overlap

(D) it shows categorical data

28. About what percent of the data falls between Q_1 and Q_3 in any box plot?

(A) 25%

(B) 50%

(C) 75%

(D) 100%

29. A data set has a range of 40 but an IQR of only 8. What does this tell you?

(A) All values are close together.

(B) Most values are clustered, but there are extreme values far from the center.

(C) The median equals the mean.

(D) The data is symmetric.

30. In a stem-and-leaf plot, the stem represents —

(A) the last digit of each data value

(B) all digits except the last digit of each data value

(C) the average of the data set

(D) the number of data values

 End of Practice Test 6 

Great job finishing the test!

 My Score

I got ___________ out of 30 questions right.

Check your answers in the **Answer Key** at the back of the book.

💡 Review any questions you missed. That's how we learn!

📊 Check Your Score Online!

Visit **ViewMath Academy** to enter your answers and see which topics you need to review. You can also explore lessons, take quizzes, track your scores, and save your progress!

viewmath.com/score/6.1.TX.21

Or go to viewmath.com/score and enter code: 6.1.TX.21

Practice Test 7

 30 Questions

---

 Before You Start 

- ✓ **Read each question carefully** before choosing your answer.
- ✓ **Show your work** on scratch paper when you need to.
- ✓ **Skip hard questions** and come back to them later.
- ✓ **Check your answers** when you're done.
- ✓ **Take your time** — there's no rush!

 You've Got This! 

Do your best and show what you know!

1. A lemonade stand sells 7 cups of lemonade **for each** 2 cups of iced tea. If they sold 21 cups of lemonade, how many cups of iced tea did they sell?

Your Answer:

2. A car uses 8 gallons of gas for 240 miles. How many miles per gallon does it get?

(A) 25

(B) 28

(C) 30

(D) 32

3. Complete the ratio table for the ratio 5 : 8.

5	8
10	?
?	32

Your Answer:

4. What is 100% of any number?

(A) 0

(B) Half the number

(C) The number itself

(D) Double the number

5. A recipe calls for 3 pints of broth. You have a 1-quart container. How many quarts is 3 pints? (1 quart = 2 pints) Will one container be enough?

Your Answer:

6. Which of the following is a **need** rather than a **want**?

   (A) A new video game

   (B) A pair of designer sunglasses

   (C) Groceries for the week

   (D) Concert tickets

7. Two friends each earn money mowing lawns.

   Ella: $y = 15x$ **Jake:** $y = 12x + 10$

   Who has a proportional relationship between lawns mowed (x) and earnings (y)? After mowing 6 lawns, who earns more?

   Your Answer:

8. Lily saves \$50 per month. Her monthly income is \$250. What is her savings rate?

   (A) 10%

   (B) 15%

   (C) 20%

   (D) 25%

9. A submarine is 250 feet below sea level. Which integer represents the submarine's position?

   (A) 250

   (B) -250

   (C) 0

   (D) -25

10. A company's quarterly result is $-\$200$. A manager says, "We are 200 dollars from break-even." Which mathematical concept did the manager use?

    (A) The opposite of -200

    (B) The absolute value $|-200| = 200$

    (C) Subtracting $200 - 0 = 200$

    (D) The reciprocal of 200

Find more at
ViewMath.com/TX-Grade6

11. In three consecutive football plays, a team gained 6 yards, lost 11 yards, and gained 3 yards. What was the team's net yardage for the three plays?

Your Answer:

12. What is $(-1) \times (-2) \times (-5) \times (-3)$?

(A) -30

(B) 30

(C) -11

(D) 11

13. Evaluate: 2×3^2

(A) 12

(B) 18

(C) 36

(D) 64

14. Which expression represents "a number r squared, minus 10"?

(A) $2r - 10$

(B) $r^2 - 10$

(C) $(r - 10)^2$

(D) $r - 10^2$

15. List all the terms in the expression $6a - 4b + 11$.

Your Answer:

16. A rectangle has length l and width 5. Which expression represents its area?

(A) $l + 5$

(B) $2l + 10$

(C) $5l$

(D) $l - 5$

17. Solve: $x + 9 = 15$

(A) $x = 24$

(B) $x = 6$

(C) $x = 4$

(D) $x = 9$

18. Which phrase matches the inequality $t \leq 30$?

(A) The temperature is more than 30 degrees

(B) The temperature is exactly 30 degrees

(C) The temperature is no more than 30 degrees

(D) The temperature is at least 30 degrees

19. A graph shows a closed circle at -1 and shading to the left. Which number is NOT a solution?

(A) -1

(B) -5

(C) -10

(D) 0

20. A triangular flower bed has a base of 4.5 m and a height of 6 m. What is its area?

Your Answer:

21. A toy box is 3 ft long, 2 ft wide, and 2 ft tall. What is the volume?

(A) $7 \ ft^3$

(B) $12 \ ft^3$

(C) $6 \ ft^3$

(D) $24 \ ft^3$

22. To find the length of a horizontal side on the coordinate plane, you subtract the:

(A) y-coordinates and take the absolute value

(B) x-coordinates and take the absolute value

(C) x-coordinate from the y-coordinate

(D) coordinates and divide by 2

Find more at
ViewMath.com/TX-Grade6

23. A rectangle has vertices $(-4, -3)$, $(6, -3)$, $(6, 2)$, and $(-4, 2)$. What is the area?

Your Answer:

24. Which net below could fold into a cube?

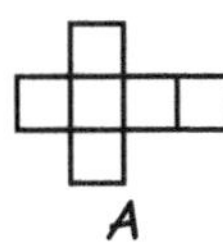
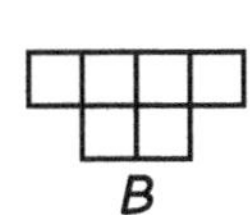
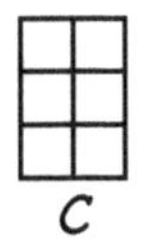

A
B
C

(A) Net A only
(B) Net B only
(C) Net C only
(D) Nets A and B

25. Data: $16, 20, 24, 28, 32, 36$. Find the median.

Your Answer:

26. Which statement about MAD is true?

(A) MAD measures the middle value of the data.
(B) MAD tells you the average distance of each value from the mean.
(C) MAD is always larger than the range.
(D) MAD equals the range divided by 2.

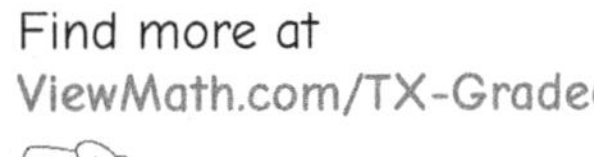

Find more at
ViewMath.com/TX-Grade6

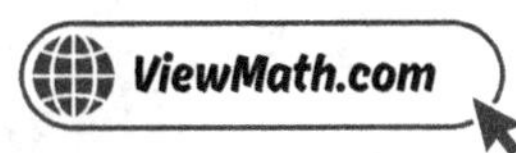

27. Look at the histogram below showing student test scores.

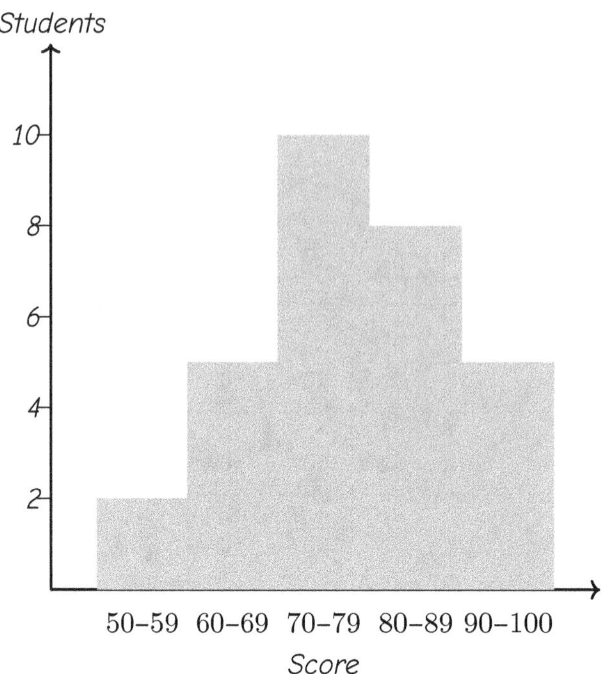

How many students took the test? What is the modal interval? What fraction of students scored 70 or higher?

Your Answer:

28. The five-number summary of a data set includes the minimum, first quartile (Q_1), median, third quartile (Q_3), and maximum. What does Q_1 represent?

(A) The lowest value in the data

(B) The middle value of the lower half of the data

(C) The average of the data

(D) The middle value of the upper half of the data

29. Which situation would make the mean a poor measure of center?

(A) A data set with all equal values

(B) A data set that is perfectly symmetric

(C) A data set with one value much larger than the rest

(D) A data set with an even number of values

Find more at
ViewMath.com/TX-Grade6

ViewMath.com

30. In the plot below, one leaf marked with "?" is missing. The median of the full data set is 37. What is the missing leaf?

Stem	Leaves
2	0 4
3	1 5 ?
4	2 6 8
5	3

Key: 2 | 0 means 20

(A) 5

(B) 6

(C) 7

(D) 9

 # End of Practice Test 7 

Great job finishing the test!

 My Score

I got _____________ out of 30 questions right.

Check your answers in the **Answer Key** at the back of the book.

Review any questions you missed. That's how we learn!

📊 Check Your Score Online!

Visit **ViewMath Academy** to enter your answers and see which topics you need to review. You can also explore lessons, take quizzes, track your scores, and save your progress!

viewmath.com/score/6.1.TX.22

Or go to viewmath.com/score and enter code: 6.1.TX.22

Practice Test 8

30 Questions

✏️ Before You Start ✏️

- ✓ **Read each question carefully** before choosing your answer.
- ✓ **Show your work** on scratch paper when you need to.
- ✓ **Skip hard questions** and come back to them later.
- ✓ **Check your answers** when you're done.
- ✓ **Take your time** — there's no rush!

 You've Got This! 

Do your best and show what you know!

1. The ratio of cats to dogs at a pet store is 3 : 4. Each part in the tape diagram represents 2 animals. How many dogs are there?

   (A) 6

   (B) 8

   (C) 3

   (D) 4

2. A family is choosing between two cell phone plans.

Plan	Lines	Monthly Cost
Family Value	4	$120
Family Plus	6	$150

Part A: Find the cost per line for each plan.

Part B: The family has 5 members. They can add an extra line to Family Value for $35, or drop a line from Family Plus for a $20 credit. What is the total cost for 5 lines under each option?

Part C: Which option is the better deal for 5 lines?

Your Answer

3. A smoothie recipe uses 2 cups of strawberries for every 3 cups of yogurt. How many cups of yogurt are needed for 10 cups of strawberries?

Your Answer

4. Which represents 35% as a fraction in simplest form?

   (A) $\dfrac{35}{10}$

   (B) $\dfrac{7}{20}$

   (C) $\dfrac{35}{50}$

   (D) $\dfrac{7}{10}$

Find more at
ViewMath.com/TX-Grade6

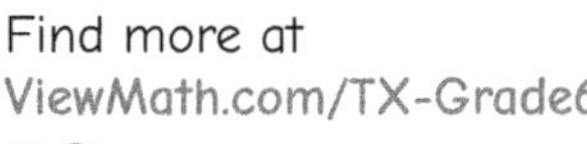

ViewMath.com

5. *A truck can carry 3 tons. A shipment weighs 5,500 pounds. Can the truck carry the shipment? (1 ton = 2,000 pounds)*

Your Answer:

6. *The table compares four payment methods.*

Method	Uses Your Own Money?	Can Charge Interest?
Cash	Yes	No
Check	Yes	No
Debit Card	Yes	No
Credit Card	No	Yes

Based on the table, which statement is true?

(A) A check borrows money from the bank

(B) A debit card can charge you interest

(C) Only a credit card involves borrowing money

(D) Cash and credit cards work the same way

7. *If a proportional relationship includes the point* $(5, 35)$, *what is the value of* y *when* $x = 8$?

(A) 48

(B) 40

(C) 56

(D) 64

8. *Two friends each start saving. Plan A saves $60 per month. Plan B saves $75 per month. After 4 months, how much **more** has Plan B saved than Plan A?*

(A) $15

(B) $30

(C) $45

(D) $60

Find more at
ViewMath.com/TX-Grade6

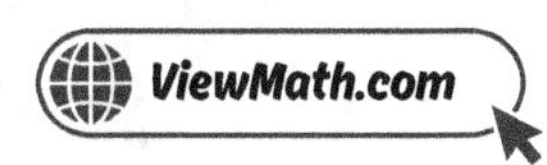

9. A city recorded these temperatures during one week: 4°F, −3°F, 0°F, −6°F, 2°F. How many of these temperatures are below zero?

(A) 1

(B) 2

(C) 3

(D) 4

10. An investor's account balance is −$600. How many dollars from zero is the balance? What deposit would bring it to exactly $0?

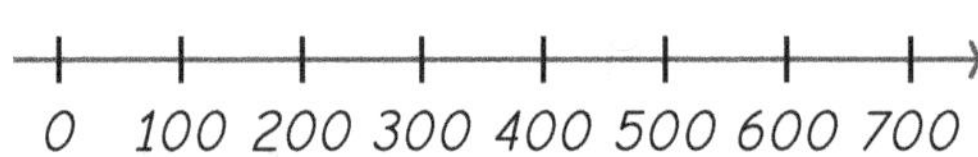

Your Answer:

11. The vertical scale below represents a thermometer. The temperature starts at 8°C and then drops 13°C.

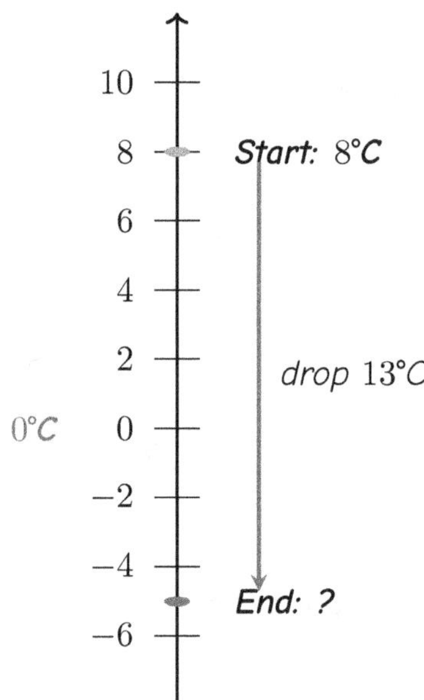

What is the temperature after the drop?

(A) −5°C

(B) 5°C

(C) −21°C

(D) 21°C

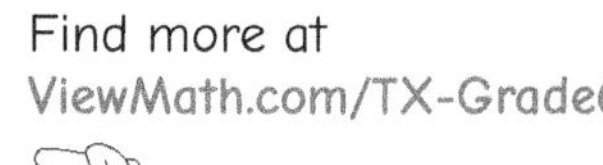
Find more at
ViewMath.com/TX-Grade6

ViewMath.com

12. What is $(-2) \times (-3) \times (-4)$?

(A) 24

(B) -9

(C) 9

(D) -24

13. Insert one pair of parentheses to make this equation true: $2 + 3 \times 4 = 20$.

Your Answer:

14. Look at the two-step diagram below. Write the expression that describes the final output when starting with the input n.

$$\boxed{\text{Input: } n} \longrightarrow \boxed{\text{Multiply by } 3} \longrightarrow \boxed{\text{Subtract } 8} \longrightarrow \boxed{\text{Output: ?}}$$

Your Answer:

15. In the expression $5(y + 8)$, name the two factors.

Your Answer:

16. A concert hall has r rows with 20 seats in each row plus 15 VIP seats in the front. Write an expression for the total number of seats.

Your Answer:

17. Solve: $\dfrac{w}{8} = 9$

Your Answer:

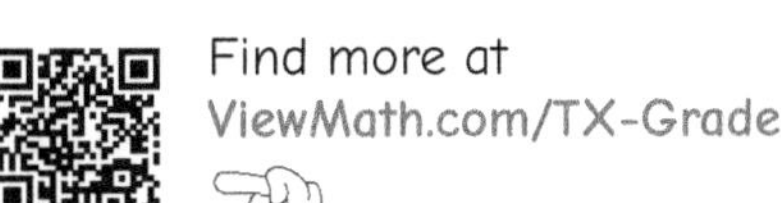
Find more at
ViewMath.com/TX-Grade6

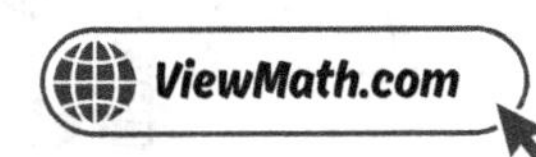
ViewMath.com

18. Which symbol correctly completes the statement? "The speed limit is 65 mph. You must drive _______ 65 mph."

(A) $>$

(B) $<$

(C) $\leq$

(D) $\geq$

19. A student says the graph of $x < 8$ and the graph of $x \leq 8$ look exactly the same. Is this correct?

Your Answer:

20. Look at the two triangles on the grid. Which triangle has a greater area?

(A) Triangle A

(B) Triangle B

(C) They have the same area.

(D) Not enough information to tell.

21. A rectangular prism has length $\frac{1}{3}$ ft, width $\frac{1}{2}$ ft, and height 6 ft. What is the volume?

Your Answer:

22. What is the distance between $(-4, 3)$ and $(6, 3)$?

(A) 2 units

(B) 10 units

(C) 6 units

(D) 7 units

23. A rectangle has vertices $(-3, 2)$, $(5, 2)$, $(5, -4)$, and $(-3, -4)$. What is the area?

(A) 24 square units

(B) 36 square units

(C) 48 square units

(D) 28 square units

24. A box is 15 in long, 6 in wide, and 4 in tall. How many square inches of cardboard are needed to make the box?

Your Answer:

25. Data: $3, 4, 5, 6, 7, 8, 60$. Find the mean and median. Which better describes a typical value?

Your Answer:

26. Which measure of spread is most affected by outliers?

(A) IQR

(B) MAD

(C) Range

(D) Median

27. A dot plot of daily steps: 3000 (1), 4000 (2), 5000 (6), 6000 (4), 7000 (2). What is the mode? What is the range?

Your Answer:

28. Data: $4, 6, 7, 8, 10, 11, 14, 15, 18, 20, 22$. What is Q_3?

(A) 14

(B) 15

(C) 18

(D) 20

29. A student says "The data set with the larger range is always more spread out than the one with the smaller range." Is this always true? Explain.

Your Answer

30. Using the stem-and-leaf plot below, how many data values are greater than 50?

Stem	Leaves
3	2 5 8
4	0 4 7 9
5	1 3 6
6	2

Key: 3 | 2 means 32

A) 3

B) 4

C) 5

D) 6

 # End of Practice Test 8 

Great job finishing the test!

 My Score

I got _____________ out of 30 questions right.

*Check your answers in the **Answer Key** at the back of the book.*

💡 *Review any questions you missed. That's how we learn!*

📊 Check Your Score Online!

Visit **ViewMath Academy** to enter your answers and see which topics you need to review. You can also explore lessons, take quizzes, track your scores, and save your progress!

viewmath.com/score/6.1.TX.23

Or go to viewmath.com/score and enter code: 6.1.TX.23

Practice Test 9

 30 Questions

Before You Start

- **Read each question carefully** before choosing your answer.
- **Show your work** on scratch paper when you need to.
- **Skip hard questions** and come back to them later.
- **Check your answers** when you're done.
- **Take your time** — there's no rush!

 You've Got This!

Do your best and show what you know!

1. The ratio of red to blue to green beads is $1 : 3 : 2$. There are 18 beads total. How many blue beads are there?

   (A) 3

   (B) 6

   (C) 9

   (D) 12

2. A 12-pack of juice boxes costs $5.40. What is the unit price per juice box?

   (A) $0.40

   (B) $0.45

   (C) $0.50

   (D) $0.55

3. Look at this ratio table. What is the missing value?

Apples	Oranges
3	5
6	?

   (A) 8

   (B) 10

   (C) 15

   (D) 11

4. A video game costs $50. Sales tax is 6%. What is the total cost?

   (A) $53

   (B) $56

   (C) $55

   (D) $52

5. Convert 3 gallons to pints. (1 gallon = 8 pints)

   (A) 16

   (B) 24

   (C) 32

   (D) 11

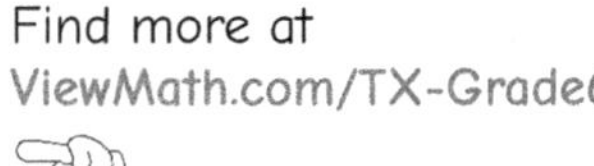
Find more at
ViewMath.com/TX-Grade6

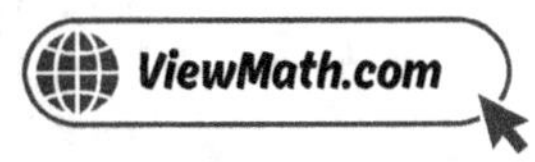

6. Which of the following is a **want** rather than a **need**?

   (A) Warm clothing in winter

   (B) Drinking water

   (C) A streaming music subscription

   (D) A visit to the doctor

7. Look at the two tables below.

Table 1: x: $1, 2, 4$ y: $3, 6, 12$ **Table 2:** x: $1, 2, 4$ y: $3, 7, 15$

Which statement is correct?

   (A) Both tables are proportional.

   (B) Only Table 1 is proportional.

   (C) Only Table 2 is proportional.

   (D) Neither table is proportional.

8. The bar chart below shows how much two friends saved each month for 5 months. Each friend started with \$0.

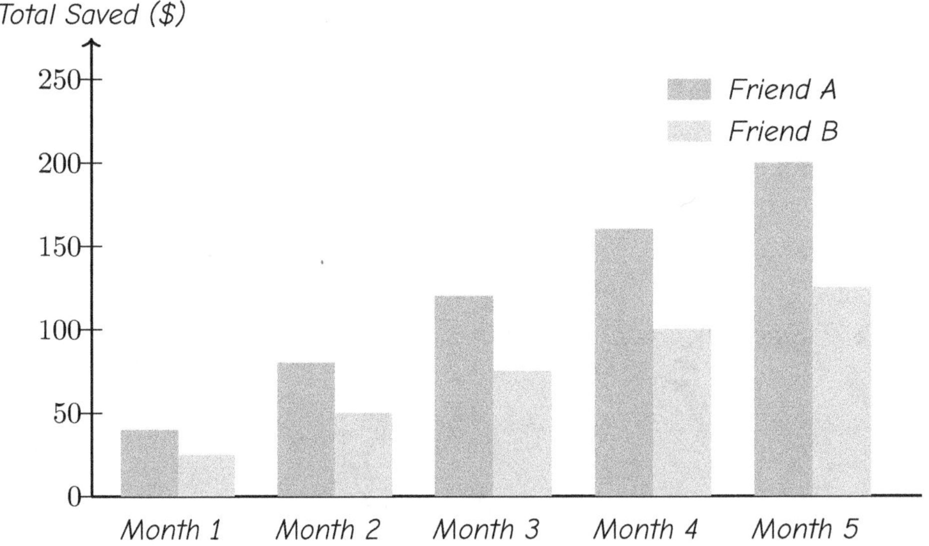

Friend A saves \$40 per month and Friend B saves \$25 per month.

Part A: After 5 months, how much more has Friend A saved than Friend B?

Part B: If Friend B deposits her 5-month total into an account earning 4% simple interest per year, how much interest will she earn in 1 year?

Your Answer:

9. A mine shaft goes 120 feet underground. Which integer describes its depth?

(A) 120

(B) -120

(C) 0

(D) -12

10. The table below shows the end-of-month account balances for five small businesses.

Business	Balance
Sunny Snacks	$-\$30$
Pet Palace	$\$45$
Book Barn	$-\$60$
Toy Town	$\$15$
Green Garden	$-\$45$

Part A: Find the absolute value of each business's balance.

Part B: Which business is farthest from break-even? Explain how you know.

Part C: Which two businesses are the same distance from break-even, even though their balances have different signs?

Your Answer:

11. What is $(-3) + (-5) + 2$?

(A) -6

(B) -10

(C) 0

(D) 4

12. What is 6×8?

(A) -48

(B) 48

(C) 14

(D) -14

13. *Evaluate:* $48 \div (4 + 4)$

    (A) 16

    (B) 6

    (C) 8

    (D) 12

14. *Which expression represents "7 less than a number y"?*

    (A) $7 - y$

    (B) $y - 7$

    (C) $7 + y$

    (D) $7y$

15. *In $2(a + 9)$, which of the following is NOT a factor?*

    (A) 2

    (B) $(a + 9)$

    (C) a

    (D) All are factors of the expression

16. *Emma earns \$8 per hour babysitting. She also got a \$15 tip. Which expression gives her total earnings for h hours?*

    (A) $8 + 15h$

    (B) $23h$

    (C) $8h + 15$

    (D) $8h - 15$

17. *Solve:* $8w = 72$

    (A) $w = 8$

    (B) $w = 64$

    (C) $w = 9$

    (D) $w = 80$

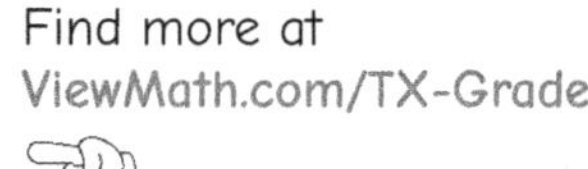

Find more at
ViewMath.com/TX-Grade6

18. The temperature stayed below 0 degrees. Which inequality represents the temperature t?

(A) $t > 0$

(B) $t < 0$

(C) $t \leq 0$

(D) $t \geq 0$

19. The pool opens when the temperature is more than $75°F$. Which describes the graph of this inequality?

(A) Closed circle at 75, shade right

(B) Open circle at 75, shade right

(C) Closed circle at 75, shade left

(D) Open circle at 75, shade left

20. A triangle has base $\frac{1}{2}$ ft and height $\frac{1}{4}$ ft. What is the area?

(A) $\frac{1}{8}$ ft^2

(B) $\frac{3}{4}$ ft^2

(C) $\frac{1}{16}$ ft^2

(D) $\frac{1}{4}$ ft^2

21. Two boxes have the same volume. Box A is $6 \times 4 \times 5$. Box B is $10 \times 3 \times h$. What is the height h of Box B?

(A) 2

(B) 4

(C) 6

(D) 8

22. What is the distance between the points $(2, 5)$ and $(2, -3)$?

(A) 2 units

(B) 5 units

(C) 8 units

(D) 3 units

23. A right triangle has vertices $(0, 0)$, $(10, 0)$, and $(10, 4)$. What is the area?

(A) 20 square units

(B) 40 square units

(C) 14 square units

(D) 7 square units

Find more at
ViewMath.com/TX-Grade6

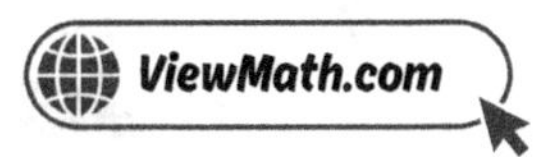

24. A cube has edges of 5 in. What is the surface area?

(A) $25\ in^2$

(B) $125\ in^2$

(C) $150\ in^2$

(D) $75\ in^2$

25. Look at the dot plot below showing the number of goals scored per game.

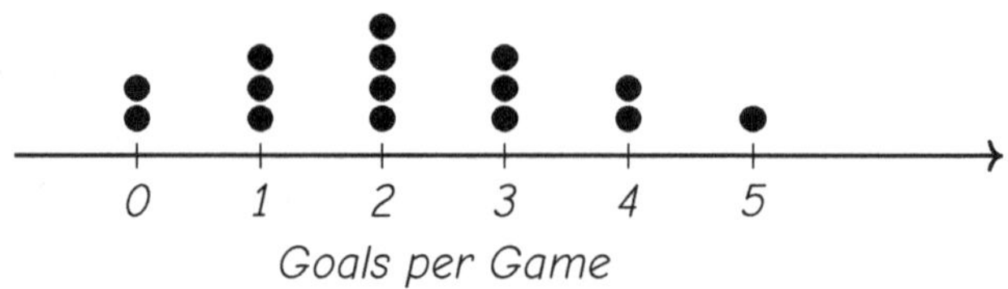

What is the median number of goals?

(A) 1

(B) 2

(C) 2.5

(D) 3

26. If every value in a data set is increased by 5, what happens to the range?

(A) It increases by 5.

(B) It decreases by 5.

(C) It stays the same.

(D) It doubles.

27. *The histogram below shows the number of pages students read last week.*

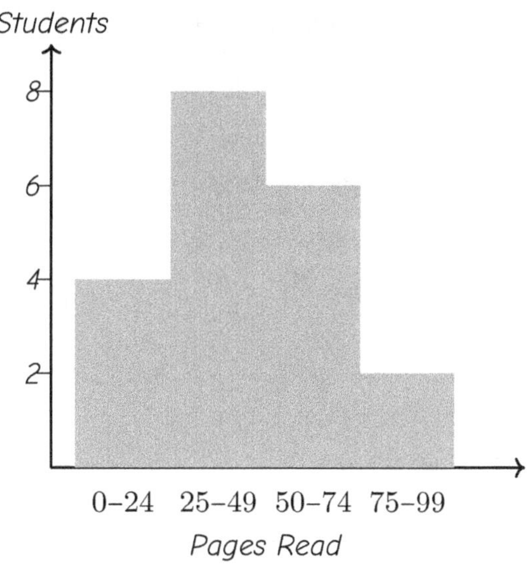

Which interval is the modal interval (contains the most students)?

(A) 0–24

(B) 25–49

(C) 50–74

(D) 75–99

28. *Team X's box plot:* min = 50, Q_1 = 60, *median* = 70, Q_3 = 80, *max* = 90. *Team Y's box plot:* min = 55, Q_1 = 68, *median* = 75, Q_3 = 82, *max* = 88. *Which team has the higher median score?*

(A) *Team X*

(B) *Team Y*

(C) *They are equal.*

(D) *Cannot be determined.*

29. *Data set X:* 10, 20, 30, 40, 50, 60, 70. *Data set Y:* 35, 37, 38, 40, 42, 43, 45. *Both have median 40. Which data set has a smaller IQR?*

(A) *Data set X*

(B) *Data set Y*

(C) *They are the same.*

(D) *Cannot be determined.*

30. *Find the mean of the data in the stem-and-leaf plot.*

Stem	Leaves
2	0 4 8
3	2 6

Key: 2 | 0 means 20

Your Answer:

 End of Practice Test 9 

Great job finishing the test!

 My Score

I got ___________ out of 30 questions right.

*Check your answers in the **Answer Key** at the back of the book.*

💡 *Review any questions you missed. That's how we learn!*

📊 Check Your Score Online!

Visit **ViewMath Academy** to enter your answers and see which topics you need to review. You can also explore lessons, take quizzes, track your scores, and save your progress!

viewmath.com/score/6.1.TX.24

Or go to viewmath.com/score and enter code: 6.1.TX.24

Practice Test 10

30 Questions

✏️ **Before You Start** ✏️

- ✔ **Read each question carefully** before choosing your answer.
- ✔ **Show your work** on scratch paper when you need to.
- ✔ **Skip hard questions** and come back to them later.
- ✔ **Check your answers** when you're done.
- ✔ **Take your time** — there's no rush!

⭐ **You've Got This!** ⭐

Do your best and show what you know!

1. The tape diagram below shows the ratio of apple juice to orange juice in a drink.

Apple: ▢▢▢
Orange: ▢▢▢▢▢

If each part equals 4 ounces, how many total ounces of drink are there?

(A) 12

(B) 20

(C) 32

(D) 8

2. The table below compares two brands of laundry detergent.

Brand	Loads	Price
Clean & Fresh	32	$9.60
Sparkle Wash	48	$12.00

Which brand costs less per load?

(A) Clean & Fresh ($0.30/load)

(B) Sparkle Wash ($0.30/load)

(C) Clean & Fresh ($0.25/load)

(D) Sparkle Wash ($0.25/load)

3. Which pair of ratios are equivalent?

(A) 2 : 3 and 4 : 9

(B) 2 : 3 and 6 : 9

(C) 2 : 3 and 8 : 9

(D) 2 : 3 and 3 : 2

4. A sweater costs $45. You have a 20% off coupon, but must pay 8% sales tax on the sale price. What is the total cost?

(A) $36.00

(B) $38.88

(C) $40.50

(D) $42.12

5. Which unit conversion uses a ratio?

(A) $5 + 12 = 17$ *inches*

(B) $\dfrac{5\ ft}{1} \times \dfrac{12\ in}{1\ ft} = 60\ in$

(C) $5 - 12 = -7$ *inches*

(D) $\dfrac{12}{5} = 2.4$ *feet*

6. The budget chart below shows how a student plans to split $500 of summer job earnings.

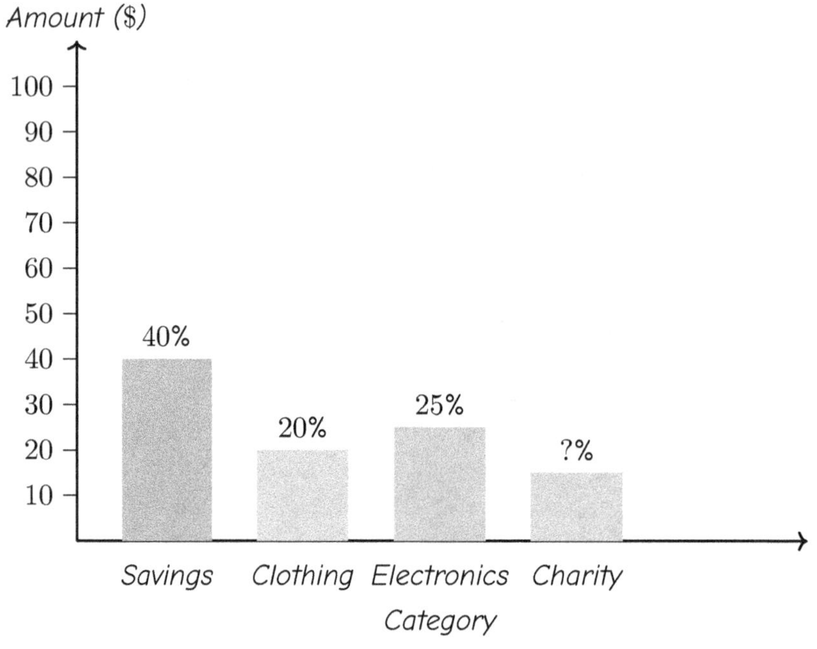

Part A: *What percent of the earnings goes to Charity?*

Part B: *How many dollars go to each of the four categories?*

Part C: *If the student decides to move 5% from Electronics to Savings, what are the new dollar amounts for those two categories?*

Your Answer:

7. A car wash charges $8 per car with no membership fee. Is this a proportional relationship? If yes, write the equation. If no, explain why not.

Your Answer:

8. Emma wants to buy a $120 skateboard. She earns $40 per week and saves 50% of her income. How many weeks will it take her to save enough money?

9. On a number line, where are negative numbers located?

(A) To the right of zero

(B) To the left of zero

(C) Above zero

(D) On top of zero

10. A number plus its opposite always equals which value?

(A) 1

(B) The number itself

(C) 0

(D) −1

11. The number line below shows three consecutive jumps starting from 0.

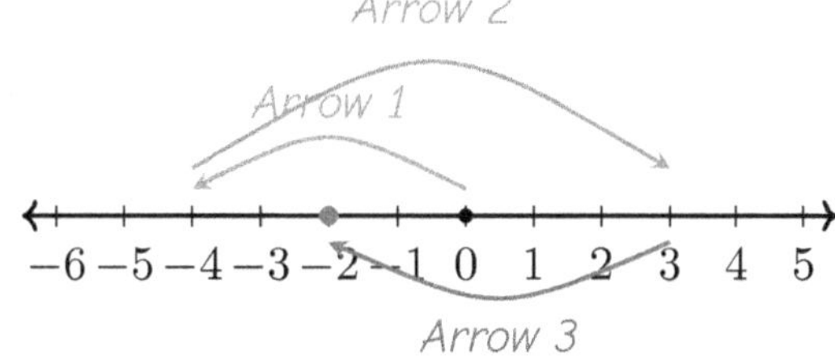

Part A: Write the integer that each arrow represents.

Part B: Write the complete addition expression shown by the three arrows.

Part C: What is the final value (the position of the blue dot)?

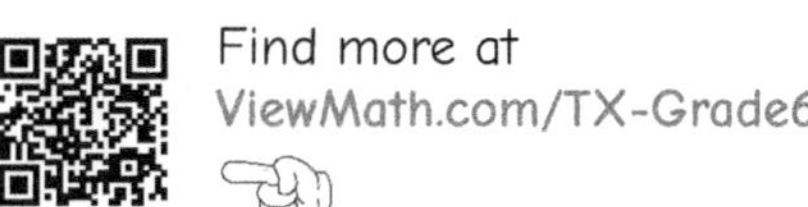

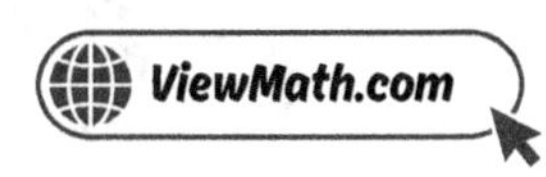

12. A submarine descends 8 feet per minute. What is the submarine's change in elevation after 7 minutes?

(A) 56 feet

(B) -15 feet

(C) 15 feet

(D) -56 feet

13. The area of each small square below is 1 square unit. Write the area using an exponent, then find the total area.

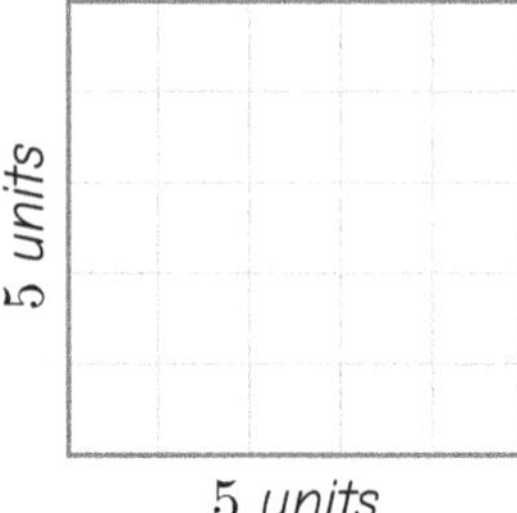

Your Answer

14. Which expression represents "6 more than half of a number n"?

(A) $6 + 2n$

(B) $\dfrac{n + 6}{2}$

(C) $\dfrac{n}{2} + 6$

(D) $n + 3$

15. Which statement is true about the expression $x + 5$?

(A) The coefficient of x is 0

(B) There are 3 terms

(C) 5 is the coefficient of x

(D) The coefficient of x is 1

16. A taxi charges \$3 plus \$2 per mile. Write an expression for the cost of a ride that is m miles long.

Your Answer

17. *Solve:* $x + 3.5 = 10$

(A) $x = 6.5$

(B) $x = 13.5$

(C) $x = 7.5$

(D) $x = 3.5$

18. Look at the elevator sign below. Write an inequality for the maximum number of people p and the maximum weight w.

ELEVATOR CAPACITY

Maximum: 12 persons

Maximum weight: 2,000 lbs

Your Answer:

19. Which inequality has a graph with an open circle at 0 and shading to the right?

(A) $x \leq 0$

(B) $x \geq 0$

(C) $x > 0$

(D) $x < 0$

20. A triangle has an area of 24 ft^2 and a base of 8 ft. What is the height?

(A) 3 ft

(B) 6 ft

(C) 16 ft

(D) 4 ft

21. What is the volume of a rectangular prism with length 5 cm, width 3 cm, and height 4 cm?

(A) 12 cm^3

(B) 60 cm^3

(C) 30 cm^3

(D) 94 cm^3

22. A triangle on the coordinate plane has vertices $(0,0)$, $(8,0)$, and $(4,6)$. What is the length of the base along the x-axis?

(A) 4 units

(B) 6 units

(C) 8 units

(D) 10 units

23. A right triangle has vertices $(-3,-2)$, $(5,-2)$, and $(-3,4)$. What is the area?

Your Answer

24. The diagram shows a rectangular prism with its dimensions. What is the area of the shaded face?

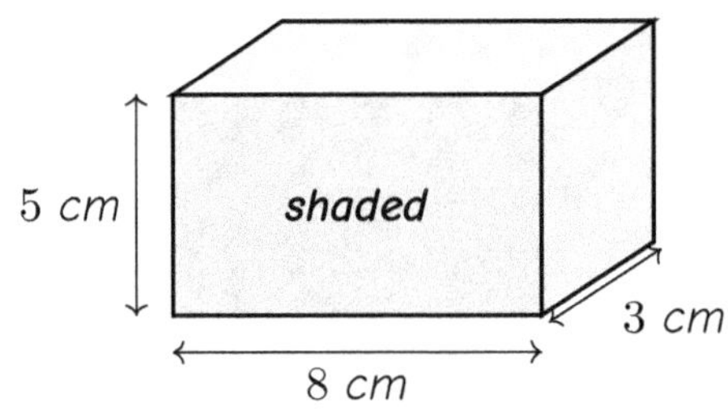

(A) $15\ cm^2$

(B) $24\ cm^2$

(C) $40\ cm^2$

(D) $120\ cm^2$

25. Four students' heights in inches are: $58, 60, 62, 64$. A fifth student who is 82 inches tall joins. What happens to the mean?

(A) It stays at 61.

(B) It increases to 65.2.

(C) It decreases.

(D) It increases to 82.

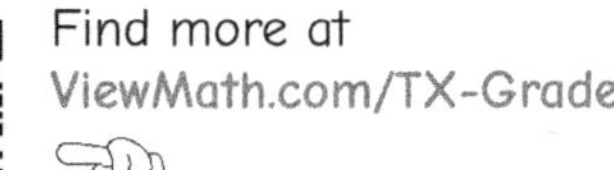
Find more at
ViewMath.com/TX-Grade6

ViewMath.com

26. Two data sets have the same mean of 50. Data set A: MAD $= 2$. Data set B: MAD $= 10$. Which data set has more variability?

Your Answer:

27. Quiz scores: $7, 8, 8, 9, 8, 10, 9, 7, 8, 9$. What is the mode?

(A) 7

(B) 8

(C) 9

(D) 10

28. The box plot below shows the commute times (in minutes) for employees at a company.

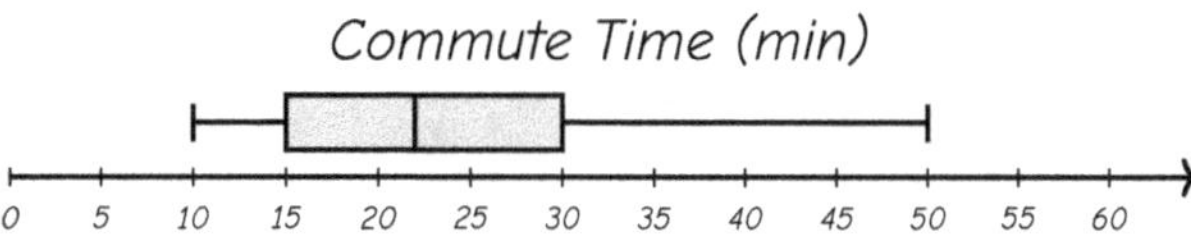

What is the IQR of the commute times?

(A) 10

(B) 15

(C) 22

(D) 40

29. You are writing a report about two data sets. Data set 1 has a symmetric shape; data set 2 is skewed. Which center measure should you use for each?

(A) Mean for both

(B) Median for both

(C) Mean for symmetric; median for skewed

(D) Mode for both

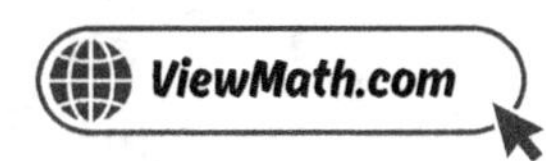

30. *Which stem has the most data values?*

Stem	Leaves
4	0 2
5	1 3 5 7 9
6	2 6 8
7	4

Key: 4 | 0 means 40

(A) 4

(B) 5

(C) 6

(D) 7

Find more at
ViewMath.com/TX-Grade6

ViewMath.com

 # End of Practice Test 10 

Great job finishing the test!

 My Score

I got __________ out of 30 questions right.

*Check your answers in the **Answer Key** at the back of the book.*

💡 *Review any questions you missed. That's how we learn!*

📊 Check Your Score Online!

Visit **ViewMath Academy** to enter your answers and see which topics you need to review. You can also explore lessons, take quizzes, track your scores, and save your progress!

viewmath.com/score/6.1.TX.25

Or go to viewmath.com/score and enter code: 6.1.TX.25

Answer Key & Explanations

Answer Key

First try each test on your own, then check your work here.

✓ Practice Test 1 — Answer Key

 5    C    C    B    C    $15    B    B    -5  32

 B

 Part A: Hour 4: -12, Hour 5: -15, Hour 6: -18. Part B: $6 \times (-3) = -18\,°F$. Part C: $20 + (-18) = 2\,°F$.

 41    C   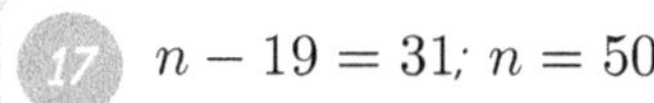 12 and k  D    $n - 19 = 31;\ n = 50$

 $n = 4$ is NOT a solution to $n > 4$ (since 4 is not greater than 4). $n = 4$ IS a solution to $n \geq 4$ (since 4 equals 4).

 C    C    C    D    B    D    C    C    B

 Range $= 75$, IQR $= 25$.   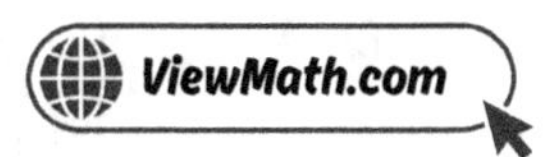 B   30 45

💡 Time to Learn! 💡

*Review the explanations below, **especially for the questions you missed**.*

Understanding why each answer is correct builds stronger problem-solving skills.

Tip: *Circle any questions you got wrong, then read their explanation carefully.*

📖 Practice Test 1 — Detailed Explanations

1 Each part $= 20 \div 4 = 5$. Vegetables $= 1 \times 5 = 5$.

2 From the graph, 2 notebooks cost \$3. Unit price: $\$3 \div 2 = \$1.50$ per notebook.

3 From the double number line, 9 scoops aligns with 15 cups of water.

4 $0.15 \times 100 = 15\%$.

5 1 yard $= 3$ feet. $7 \times 3 = 21$ feet.

6 Used: $40\% + 25\% + 10\% = 75\%$. Left: $100\% - 75\% = 25\%$. Amount: $0.25 \times \$60 = \$15$.

7 $k = \frac{3}{1} = 3$. When $x = 4$: $y = 3 \times 4 = 12$. The line passes through the origin and $\frac{y}{x} = 3$ for every pair, so it is proportional.

8 $I = P \times r \times t = 500 \times 0.04 \times 2 = \$40$.

9 The signs are different. Subtract the absolute values: $9 - 4 = 5$. The number with the larger absolute value is -9 (negative), so the answer is -5.

10 $|-32| = 32$ because -32 is 32 units from zero. Absolute value measures distance from zero, which is always positive or zero.

11 Starting at 5, move 9 units to the left to reach -4. So $5 - 9 = -4$.

12 Each hour adds -3. Hour 4: $-9 + (-3) = -12$. Hour 5: $-12 + (-3) = -15$. Hour 6: $-15 + (-3) = -18$. Multiplication: $6 \times (-3) = -18$. Starting at $20\,°F$: $20 + (-18) = 2\,°F$.

Find more at
ViewMath.com/TX-Grade6

ViewMath.com

13 $5^2 = 25$ and $4^2 = 16$. Sum: $25 + 16 = 41$.

14 "Plus" means addition: $n + 8$.

15 $12k = 12 \times k$. The factors are 12 and k.

16 Start with \$50 and subtract what you spend: $50 - d$.

17 Add 19: $n = 31 + 19 = 50$.

18 $>$ does not include equality. $\geq$ includes equality.

19 $0 > -5$ is true. Since 0 is in the shaded region (right of -5), it confirms the graph.

20 One sail: $\frac{1}{2} \times 3 \times 7 = 10.5 \ m^2$. Two sails: $10.5 \times 2 = 21 \ m^2$.

21 Volume measures 3-dimensional space, so it uses cubic units like cm^3, in^3, or ft^3.

22 $|4 - (-5)| = |9| = 9$ units.

23 Base $= |8 - 2| = 6$. Height $= |7 - 1| = 6$. Area $= \frac{1}{2} \times 6 \times 6 = 18$ square units.

24 $SA = 6 \times 3^2 = 6 \times 9 = 54 \ cm^2$.

25 Even number of values. The two middle values are 30 and 40. Median $= (30 + 40) \div 2 = 35$.

26 $MAD = (4 + 2 + 0 + 2 + 4) \div 5 = 12 \div 5 = 2.4$.

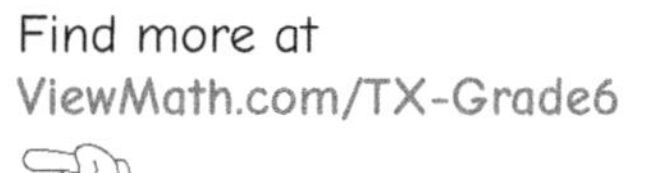
Find more at
ViewMath.com/TX-Grade6

27. *The 10–19 bar has height 10, the tallest bar.*

28. *Range $= 95 - 20 = 75$. IQR $= 60 - 35 = 25$.*

29. *School B's IQR (45) is larger than School A's (30), indicating more variability in the middle 50% of scores.*

30. *The value 45 appears 3 times (three leaves of 5 on stem 4). No other value appears more often. The mode is 45.*

✅ Practice Test 2 — Answer Key

1. C

2. *Part A: SunPower \$0.12/kWh, BrightLight \$0.11/kWh. Part B: SunPower \$120, BrightLight \$110. Part C: BrightLi*

3. *Part A: 2 : 1; Part B: 4 cups of sugar*
4. C
5. C
6. B
7. B
8. 27.5%

9. C
10. 12
11. C
12. -20
13. $3^3 = 27$ *unit cubes*
14. $12h + 20$
15. A
16. B

17. B
18. C
19. $x \le -4$
20. D
21. B
22. 26 *units*
23. B
24. 5 ft
25. B

26. Range $= 7$, IQR $= 2$
27. C
28. C
29. A
30. B

💡 Time to Learn! 💡

*Review the explanations below, **especially for the questions you missed.***

Understanding why each answer is correct builds stronger problem-solving skills.

Tip: *Circle any questions you got wrong, then read their explanation carefully.*

Find more at
ViewMath.com/TX-Grade6

📖 *Practice Test 2 — Detailed Explanations*

1 *Girls have 6 parts = 24, so each part = 24 ÷ 6 = 4. Boys have 4 parts = 4 × 4 = 16.*

2 *A: $96 ÷ 800 = $0.12/kWh; $132 ÷ 1,200 = $0.11/kWh. B: 1,000 × $0.12 = $120; 1,000 × $0.11 = $110. C: BrightLight saves $10/month.*

3 *Part A: From the graph, $(2,1)$ shows 2 cups of flour for every 1 cup of sugar. Part B: Following the pattern, $(8,4)$, so 4 cups of sugar.*

4 *$0.20 × $25 = $5.*

5 *$500 ÷ 1,000 = 0.5$ kg.*

6 *Saving regularly builds a fund for emergencies and future goals like college, a car, or unexpected bills.*

7 *A proportional relationship has the form $y = kx$ with no added or subtracted constant. Only $y = 7x$ fits this form.*

8 *Total expenses = $960 + $320 + $480 + $240 + $320 = $2,320. Savings = $3,200 − $2,320 = $880. Rate $= \dfrac{880}{3,200} = 0.275 = 27.5\%$.*

9 *A deposit adds money, which is a positive change. Debt, below sea level, and below zero are all negative.*

10 *Since $a + b = 0$, b must be the opposite of a. The opposite of −12 is 12, so $b = 12$. Check: −12 + 12 = 0. ✓*

11 *Dropping means subtracting: $5 − 12 = 5 + (−12)$. Different signs: $12 − 5 = 7$. Since 12 is larger and negative, the answer is −7°C.*

Find more at
ViewMath.com/TX-Grade6

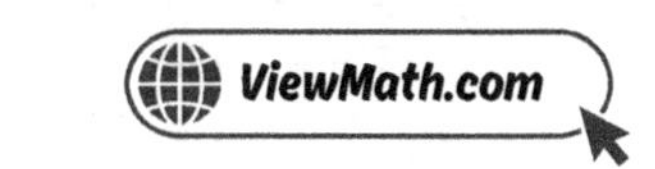

12 Different signs → negative. $120 \div 6 = 20$, so $(-120) \div 6 = -20$.

13 Each edge is 3 units. The cube needs $3 \times 3 \times 3 = 3^3 = 27$ unit cubes.

14 Hourly pay is $12h$. Plus the bonus: $12h + 20$.

15 The coefficient is the number in front of the variable. In $3y$, the coefficient is 3.

16 A square has 4 equal sides, so its perimeter is 4 times the side length: $P = 4s$.

17 Subtract 15: $k = 32 - 15 = 17$.

18 $n \leq 8$ means n is 8 or less. $8 \leq 8$ is true. 8.5, 9, and 10 are all greater than 8.

19 Closed circle means $\leq$ or $\geq$. Shading left means less than or equal to: $x \leq -4$.

20 $A = \frac{1}{2} \times 14 \times 8 = 56\ m^2$.

21 Total volume $= 5 \times 3 \times 2 = 30\ m^3$. Half-full: $30 \div 2 = 15\ m^3$.

22 Length $= |2 - (-6)| = 8$. Width $= |2 - (-3)| = 5$. Perimeter $= 2(8) + 2(5) = 26$ units.

23 Decompose irregular shapes into simpler shapes (rectangles and triangles) whose areas you can compute, then add them.

24 $2(2)(3) + 2(2)(h) + 2(3)(h) = 62$. $12 + 4h + 6h = 62$. $10h = 50$. $h = 5$ ft.

25 Mean $= (70 + 80 + 90 + 85 + 75) \div 5 = 400 \div 5 = 80$.

Find more at
ViewMath.com/TX-Grade6

26. Data: $3, 4, 4, 5, 5, 5, 6, 6, 6, 6, 7, 7, 8, 10$. 14 values. Range $= 10 - 3 = 7$. Median $= (6 + 6) \div 2 = 6$. Lower half: $3, 4, 4, 5, 5, 5, 6$, $Q1 = 5$. Upper half: $6, 6, 7, 7, 8, 10...$ wait, let me recount. 14 values: lower 7: $3, 4, 4, 5, 5, 5, 6$ ⮕ $Q1 = 5$. Upper 7: $6, 6, 6, 7, 7, 8, 10$ ⮕ $Q3 = 7$. IQR $= 7 - 5 = 2$.

27. The mode is the value with the highest frequency — it appears more often than any other value.

28. The five-number summary gives min, Q_1, median, Q_3, max — enough for a box plot. It does not tell you how many values are in the data set.

29. Class 1's scores cluster tightly (75–85), giving a small IQR despite the outlier at 40. Class 2's even spread gives a large IQR. Class 1 is more consistent.

30. Class A scores: $62, 65, 68, 71, 74, 77, 80, 86$ (8 values). Median $= (71 + 74) \div 2 = 72.5$. Class B scores: $63, 66, 70, 72, 75, 79, 81, 84, 92$ (9 values). Median $= 75$ (the 5th value). Class B has the higher median.

✅ Practice Test 3 — Answer Key

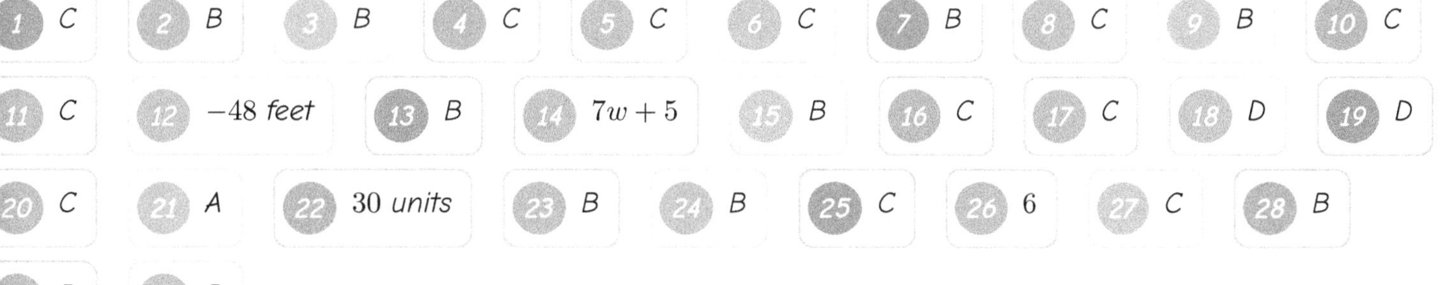

1	C	2	B	3	B	4	C	5	C	6	C	7	B	8	C	9	B	10	C
11	C	12	-48 feet	13	B	14	$7w + 5$	15	B	16	C	17	C	18	D	19	D
20	C	21	A	22	30 units	23	B	24	B	25	C	26	6	27	C	28	B
29	B	30	B																

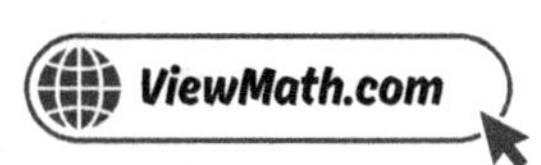

💡 Time to Learn! 💡

*Review the explanations below, **especially for the questions you missed**.*

Understanding why each answer is correct builds stronger problem-solving skills.

Tip: *Circle any questions you got wrong, then read their explanation carefully.*

📖 Practice Test 3 — Detailed Explanations

1 *The question asks flour to eggs. Flour = 3, eggs = 2. The ratio is 3 : 2.*

2 *Brand A: $5.10 \div 6 = $0.85. Brand B: $6.00 \div 8 = $0.75. Brand B is cheaper per bar.*

3 *$7 \times 3 = 21$ bananas, so $2 \times 3 = $6.*

4 *Discount: $0.15 \times $80 = $12. Sale price: $80 - $12 = $68.*

5 *$156 \div 12 = 13$ feet.*

6 *25% of $120 = 0.25 \times 120 = $30.*

7 *$k = \frac{y}{x} = \frac{10}{4} = 2.5$. The equation is $y = 2.5x$.*

8 *Savings = Income − Expenses = $600 − $450 = $150.*

9 *She starts with $15 (positive) and spends $20 (negative change). The expression is $15 + (-20)$. Different signs: $20 - 15 = 5$, and 20 is larger, so the result is -5. Her balance is $-$5.*

Find more at
ViewMath.com/TX-Grade6

10 Find each absolute value: $|-40| = 40$, $|30| = 30$, $|-55| = 55$, $|10| = 10$. The largest is 55, so Month 3 ($-\$55$) was farthest from $\$0$.

11 Rising temperature means addition: $-8 + 15$. Different signs: $15 - 8 = 7$. Since 15 is larger and positive, the answer is $7°F$.

12 Descending 4 feet $= -4$ per second. Over 12 seconds: $12 \times (-4) = -48$ feet.

13 Any number raised to the power of 0 equals 1.

14 The product of 7 and w is $7w$. Increased by 5 gives $7w + 5$.

15 $\frac{n}{5} = \frac{1}{5} \times n$. The factors are $\frac{1}{5}$ and n.

16 $50 + 40(3) = 50 + 120 = 170$ dollars.

17 Step 1 is correct (multiply to undo division), but Step 2 should be $n = 8 \times 5 = 40$, not $8 + 5 = 13$.

18 "No more than 200" means 200 or fewer. $c \leq 200$.

19 $\leq$ includes -1 (closed circle). Less than -1 is to the left (shade left).

20 $A = \frac{1}{2} \times 5 \times 12 = 30$ in^2.

21 $V = \frac{3}{4} \times 2 \times 4 = \frac{3}{4} \times 8 = 6$ ft^3.

22 Length $= |4 - (-5)| = 9$. Width $= |4 - (-2)| = 6$. Perimeter $= 2(9) + 2(6) = 30$ units.

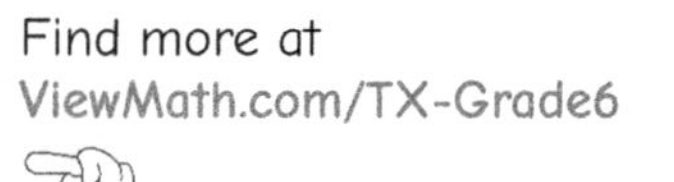

Find more at
ViewMath.com/TX-Grade6

23. $Base = |5 - 1| = 4.$ $Height = |9 - 1| = 8.$ $Area = \frac{1}{2} \times 4 \times 8 = 16$ square units.

24. One block: $2(3)(2) + 2(3)(4) + 2(2)(4) = 12 + 24 + 16 = 52$ ft^2. Two blocks: $52 \times 2 = 104$ ft^2.

25. $Total = 15 \times 6 = 90.$ Remove 9: $90 - 9 = 81.$ New mean $= 81 \div 5 = 16.2.$

26. Distances from 60: $10, 5, 0, 5, 10.$ $MAD = (10 + 5 + 0 + 5 + 10) \div 5 = 30 \div 5 = 6.$

27. Every value is 5, so 5 is the value that appears most often. The mode is 5.

28. Lower half: $10, 12, 14, 15.$ Median of lower half $= (12 + 14)/2 = 13.$ So $Q_1 = 13.$

29. Team 1 range $= 68 - 60 = 8.$ Team 2 range $= 72 - 58 = 14.$ Team 2 has the greater range.

30. The stem 11 combined with leaf 5 gives the three-digit number 115.

Practice Test 4 — Answer Key

1. B

2. Labor: $75/hour. Total per hour: $350 \div 4 = $87.50/hour.

3. C

4. Yes. $95 at 10% off $= $85.50.$ $105 at 20% off $= $84.$ Adding the item saves $1.50.

5. C

6. D

7. B

8. $90

9. A

10. B

11. C

12. 60

13. C

14. $\frac{t}{4} - 1$

15. A

16. A

17. C

18. C

19. B

20. C

21. A

22. C

23. B

24. 6 cm

25. B

26. C

27. B

28. C

29. Fertilizer A: min $= 8$, $Q_1 = 15$, med $= 24$, $Q_3 = 32$, max $= 42$. Range $= 34$, IQR $= 17$. Fertilizer B: min $= 18$, Q_1

Find more at
ViewMath.com/TX-Grade6

30. B

> ## 💡 Time to Learn! 💡
>
> Review the explanations below, **especially for the questions you missed**.
>
> Understanding why each answer is correct builds stronger problem-solving skills.
>
> **Tip:** Circle any questions you got wrong, then read their explanation carefully.

📖 Practice Test 4 — Detailed Explanations

1. "3 pencils for every 1 eraser" means pencils to erasers is $3:1$.

2. Labor rate: $\$300 \div 4 = \$75/hr$. Total cost: $\$300 + \$50 = \$350$. Total per hour: $\$350 \div 4 = \$87.50/hr$.

3. The ratio is $4:10 = 2:5$. Row 3 should be $12:30$ (since $4 \times 3 = 12$ and $10 \times 3 = 30$), but it shows $12:25$.

4. At $\$95$: 10% off gives $\$95 - \$9.50 = \$85.50$. At $\$105$: 20% off gives $\$105 - \$21 = \$84$. $\$84 < \$85.50$, so yes.

5. 1 foot $= 12$ inches. $5 \times 12 = 60$ inches.

6. A debit card takes money directly from your bank account, unlike a credit card which borrows money.

7. $y = \frac{3}{4}x$. Check each: $(3,4)$: $\frac{3}{4} \times 3 = 2.25 \neq 4$. $(8,6)$: $\frac{3}{4} \times 8 = 6$. ✓ $(5,4)$: $\frac{3}{4} \times 5 = 3.75 \neq 4$. $(6,8)$: $\frac{3}{4} \times 6 = 4.5 \neq 8$. Only $(8,6)$ works.

8. $I = P \times r \times t = 600 \times 0.05 \times 3 = \$90$.

Find more at
ViewMath.com/TX-Grade6

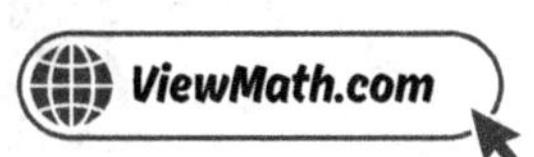

9 When both numbers have the same sign, add their absolute values and keep the sign. $4 + 3 = 7$, and both are negative, so the answer is -7.

10 Absolute value is the distance from zero, which is always positive or zero. $|-7| = 7$ because -7 is 7 units from zero on the number line.

11 Descending means subtracting: $120 - 185 = 120 + (-185)$. Different signs: $185 - 120 = 65$. Since 185 is larger and negative, the answer is -65 feet (below sea level).

12 $(-4) \times (-5) = 20$ (same signs $\rightarrow$ positive). Then $20 \times 3 = 60$. Two negative factors (even) $\rightarrow$ positive result.

13 $3^4 = 3 \times 3 \times 3 \times 3 = 81$. Each value is 3 times the previous one: $27 \times 3 = 81$.

14 Divide t by 4 to get $\frac{t}{4}$, then subtract 1: $\frac{t}{4} - 1$.

15 $7k$ is a single term. There are no $+$ or $-$ signs separating other pieces.

16 New fencing: top (12) + right (w) + bottom $(12) = 12 + w + 12 = 24 + w$.

17 Divide by 12: $p = 84 \div 12 = 7$.

18 "More than \$25" means greater than 25: $d > 25$.

19 $x < -3$: $-4 < -3$ is true (shaded), but $-2 < -3$ is false (not shaded).

20 $A = \frac{1}{2} \times 20 \times 7 = 70$ in^2.

21 Base area $= 10 \times 4 = 40$. Height $= 120 \div 40 = 3$ cm.

Find more at
ViewMath.com/TX-Grade6

22 $Length = |3 - (-5)| = 8.$ $Width = |4 - (-2)| = 6.$ $Perimeter = 2(8) + 2(6) = 28$ units.

23 $Base = 9 - 1 = 8.$ $Height = 7 - 1 = 6.$ $Area = \frac{1}{2} \times 8 \times 6 = 24$ square units.

24 $6s^2 = 216$, so $s^2 = 36$, giving $s = 6$ cm.

25 With an even number of values, the median is the average of the two middle values: $(18 + 20) \div 2 = 19$.

26 IQR measures the spread of the middle 50%. A smaller IQR (4) means those values are closer together.

27 The size of the data set and whether it is numerical or categorical determine the best display. Categorical ⊠ frequency table/bar graph. Small numerical ⊠ dot plot. Large numerical ⊠ histogram.

28 When the median is closer to Q_1, the right half of the box is wider, meaning the data is stretched to the right — skewed right.

29 Fertilizer B's higher median shows its typical plant is taller. Its smaller IQR and range show the plant heights are less spread out. Fertilizer B is the better choice if you want tall, consistent growth.

30 Stem 3: leaves $2, 5$ (2 leaves). Stem 4: leaves $1, 4, 8$ (3 leaves). Stem 5: leaves $3, 7$ (2 leaves). Stem 4 has the most.

✔ Practice Test 5 — Answer Key

1 10   **2** A   **3** C   **4** C   **5** C   **6** \$75   **7** D   **8** C   **9** C

10 $|-14| + |-9| = 14 + 9 = 23$; Alex's accounts are a combined \$23 from zero.   **11** -37 **12** B

13 4   **14** D   **15** C   **16** $100 - 8h$ **17** D   **18** C   **19** C   **20** B   **21** $60\ cm^3$

 B  A  B  B  B  Total = 30, Mode = Pizza  B

 Group X: median = $10, range = $15, IQR = $7. Group Y: median = $11, range = $3, IQR = $2.  B

💡 Time to Learn! 💡

*Review the explanations below, **especially for the questions you missed**.*

Understanding why each answer is correct builds stronger problem-solving skills.

Tip: *Circle any questions you got wrong, then read their explanation carefully.*

📖 Practice Test 5 — Detailed Explanations

1. Total parts $= 3 + 2 = 5$. Each part $= 25 \div 5 = 5$. Swimming $= 2 \times 5 = 10$.

2. Service A per month: $\$108 \div 12 = \$9$. Service B is $\$10$ per month. $\$9 < \$10$, so Service A is cheaper.

3. $4 \times 2.5 = 10$ loaves, so $6 \times 2.5 = 15$ cups. Alternatively, unit rate: $6 \div 4 = 1.5$ cups per loaf, then $1.5 \times 10 = 15$.

4. Savings $= \$40 - \$30 = \$10$. $\$10 \div \$40 = 0.25 = 25\%$.

5. 1 km $= 1{,}000$ m. $4.5 \times 1{,}000 = 4{,}500$ m.

6. $I = 250 \times 0.15 \times 2 = \$75$.

7. A proportional relationship has the form $y = kx$. When $x = 0$, $y = 0$, so the line always passes through the origin $(0, 0)$.

Find more at
ViewMath.com/TX-Grade6

8 *Savings percent* $= 100\% - 35\% - 10\% - 20\% - 15\% = 20\%$. *Before:* $0.20 \times \$1{,}000 = \$200$. *After:* $0.20 \times \$1{,}200 = \$240$. *Difference* $= \$240 - \$200 = \$40$.

9 *The signs are different. Subtract:* $8 - 5 = 3$. *The number with the larger absolute value is* -8 *(negative), so the answer is* -3.

10 $|-14| = 14$ *and* $|-9| = 9$. *Together,* $14 + 9 = 23$. *This means the total distance of both accounts from* $\$0$ *is* $\$23$, *which is the total amount Alex owes.*

11 *Same signs: add absolute values* $14 + 23 = 37$ *and keep the common (negative) sign.* $(-14) + (-23) = -37$.

12 *Positive* $\times$ *negative* $=$ *negative.* $9 \times 6 = 54$, *so* $9 \times (-6) = -54$.

13 *Parentheses:* $12 - 4 = 8$. *Exponent:* $8^2 = 64$. *Divide:* $64 \div 16 = 4$.

14 *Giving away 6 means subtracting:* $x - 6$.

15 $9x = 9 \times x$. *The factors of this term are 9 and* x.

16 *Start at 100 and subtract 8 per hour.* $100 - 8h$.

17 *Division is undone by multiplication. Multiply both sides by 3:* $n = 12 \times 3 = 36$.

18 $n \geq 12$ *means* n *is 12 or more, which is "at least 12."*

19 $-3 \geq -3$ *is true (they are equal). The other values are less than* -3.

20 $P = \frac{1}{2} \times 10 \times 6 = 30$. $Q = \frac{1}{2} \times 10 \times 8 = 40$. *Difference* $= 40 - 30 = 10$ *cm*^{2}.

Find more at
ViewMath.com/TX-Grade6

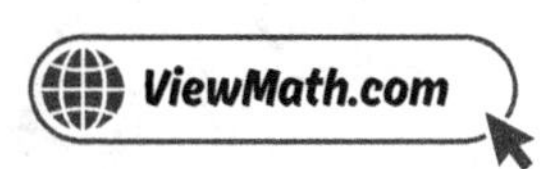

21 $V = 2.5 \times 4 \times 6 = 60 \ cm^3$.

22 Horizontal side: $|7 - 1| = 6$. Vertical side: $|2 - (-4)| = 6$. Both sides are 6 units, so it is a square. The side length is 6.

23 Length $= |6 - (-3)| = 9$. Width $= 54 \div 9 = 6$ units.

24 $SA = 2(6)(4) + 2(6)(3) + 2(4)(3) = 48 + 36 + 24 = 108 \ cm^2$.

25 The mean uses every value in its calculation, so one extreme value changes it significantly. The median depends only on the middle position.

26 Distances from 24: $4, 2, 0, 2, 4$. $MAD = (4 + 2 + 0 + 2 + 4) \div 5 = 12 \div 5 = 2.4$

27 $12 + 8 + 4 + 6 = 30$. Pizza has the highest frequency (12), so it is the mode.

28 Each whisker represents approximately 25% of the data. A longer whisker means those values are more spread out, not that there are more of them.

29 Group X: $Q_1 = 8$, $Q_3 = 15$, $IQR = 7$. Group Y: $Q_1 = 10$, $Q_3 = 12$, $IQR = 2$. Group Y gets a slightly higher typical allowance (\$11 vs. \$10) and allowances are much more consistent (IQR 2 vs. 7, range 3 vs. 15).

30 The leaves for stem 5 should be written 1 3 7 9, not 3 1 7 9. Leaves must always go from least to greatest.

✅ Practice Test 6 — Answer Key

1 Bananas to yogurt: $3 : 4$; Yogurt to bananas: $4 : 3$ **2** B   **3** B   **4** C   **5** C   **6** D

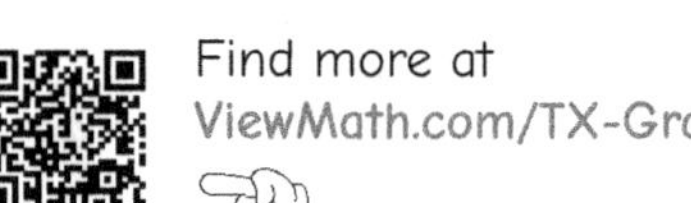

Find more at
ViewMath.com/TX-Grade6

 B  B  B  B  C  B  8  D  B  D

 $k = 9$  6, 7, and 10 (or any three numbers that are 6 or greater)    B    $\frac{1}{4}$ ft²    B

 12 units    B    C    D    A    B    B    B    B

💡 Time to Learn! 💡

Review the explanations below, **especially for the questions you missed.**

Understanding why each answer is correct builds stronger problem-solving skills.

Tip: Circle any questions you got wrong, then read their explanation carefully.

📖 Practice Test 6 — Detailed Explanations

1. "3 bananas for every 4 cups of yogurt" gives $3 : 4$. Flip the order for yogurt to bananas: $4 : 3$.

2. $\$45 \div 3 = \$15$ per month.

3. $2 \times 8 = 16$ students, so $3 \times 8 = 24$ pencils.

4. Discount: $0.25 \times \$60 = \$15$. Sale price: $\$60 - \$15 = \$45$.

5. $1\ km = 1{,}000\ m = 100{,}000\ cm$. $10 \times 100{,}000 = 1{,}000{,}000\ cm$.

6. Monthly savings: $0.15 \times \$200 = \$30$. In 6 months: $\$30 \times 6 = \$180$.

7. The line passes through the origin, so it is proportional: $y = kx$. At $(1, 6)$: $k = \frac{6}{1} = 6$. The equation is $y = 6x$.

8 Weekly savings $= 0.25 \times \$120 = \$30$. Total after 6 weeks $= \$30 \times 6 = \$180$. Interest $= 180 \times 0.03 \times 1 = \$5.40$.

9 A negative temperature means below zero. $-15°F$ means 15 degrees below zero.

10 Find each absolute value: $|3| = 3$, $|-8| = 8$, $|-1| = 1$, $|5| = 5$. The greatest absolute value is 8, which belongs to -8.

11 Subtracting a negative is the same as adding a positive: $a - (-b) = a + b$. For example, $3 - (-4) = 3 + 4 = 7$.

12 When two factors have the same sign, the product is always positive. Negative $\times$ negative $=$ positive.

13 Parentheses: $3 + 2 = 5$. Exponent: $2^2 = 4$. Divide: $60 \div 5 = 12$. Subtract: $12 - 4 = 8$.

14 "Divide 20 by p" puts 20 first: $20 \div p$.

15 The term d means $1 \cdot d$, so the coefficient is 1. (The subtraction sign belongs to the operation, not the coefficient of d itself in the original expression.)

16 Splitting $14 equally among f friends means dividing: $14 \div f$.

17 $k = 54 \div 6 = 9$. Check: $6(9) = 54$ ✓

18 Any number greater than or equal to 6 is a solution.

19 Both shade to the right. The only difference is the circle at 5: open for $>$, closed for $\geq$.

20 $A = \frac{1}{2} \times \frac{3}{4} \times \frac{2}{3} = \frac{1}{2} \times \frac{6}{12} = \frac{1}{2} \times \frac{1}{2} = \frac{1}{4}$ ft^2.

Find more at
ViewMath.com/TX-Grade6

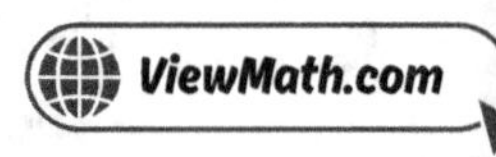

21. $V = l \times w \times h$. If h becomes $2h$, then $V_{new} = l \times w \times 2h = 2(lwh)$, so the volume doubles.

22. $|8 - (-4)| = |12| = 12$ units.

23. Width $= 56 \div 8 = 7$ units.

24. A rectangular prism has 6 faces: 3 pairs of opposite rectangles.

25. Mean $=$ sum $\div 5 = 20$, so sum $= 20 \times 5 = 100$.

26. Team A range: $68 - 60 = 8$. Team B range: $88 - 40 = 48$. Team A is much more consistent.

27. Histogram bins are consecutive numerical intervals with no values left out, so bars sit directly next to each other.

28. The box contains the middle 50% of the data (from the 25th to the 75th percentile).

29. A small IQR means the middle 50% is tightly packed. A large range means the minimum and maximum are far apart, indicating extreme values stretch the data.

30. In a stem-and-leaf plot the stem is formed by all digits except the last digit, and the leaf is the last digit.

☑ Practice Test 7 — Answer Key

1. 6

2. C

3. 16 and 20

4. C

5. 1.5 quarts; No, one 1-quart container is not enough.

6. C

Find more at
ViewMath.com/TX-Grade6

7 Ella has a proportional relationship. After 6 lawns, Ella earns $90 and Jake earns $82, so Ella earns more.

8 C   **9** B   **10** B   **11** −2 yards (a net loss of 2 yards)   **12** B   **13** B   **14** B

15 6a, 4b, and 11   **16** C   **17** B   **18** C   **19** D   **20** 13.5 m^2 **21** B   **22** B

23 50 square units   **24** A   **25** 26   **26** B

27 30 students. Modal interval: 70–79. Fraction ≥ 70: $\frac{23}{30}$.   **28** B   **29** C   **30** C

💡 Time to Learn! 💡

Review the explanations below, **especially for the questions you missed.**

Understanding why each answer is correct builds stronger problem-solving skills.

Tip: Circle any questions you got wrong, then read their explanation carefully.

📖 Practice Test 7 — Detailed Explanations

1 $21 \div 7 = 3$, so multiply both by 3: iced tea $= 2 \times 3 = 6$.

2 $240 \div 8 = 30$ miles per gallon.

3 Row 2: $5 \times 2 = 10$, so $8 \times 2 = 16$. Row 3: $8 \times 4 = 32$, so $5 \times 4 = 20$.

4 100% means "the whole thing." 100% of $n = 1 \times n = n$.

5 $3 \div 2 = 1.5$ quarts. Since $1.5 > 1$, one 1-quart container is not enough.

6   Groceries are food, which is a basic need. The other choices are things you might enjoy but can live without.

7   Ella: $y = 15x$ is proportional ($k = 15$). Jake: $y = 12x + 10$ is non-proportional ($b = 10 \neq 0$). For $x = 6$: Ella earns $15 \times 6 = 90$; Jake earns $12 \times 6 + 10 = 82$. Ella earns more.

8   Savings rate $= \dfrac{\$50}{\$250} = 0.20 = 20\%$.

9   Below sea level is represented by a negative number. 250 feet below sea level is -250.

10   The manager found the distance from zero, which is the absolute value. $|-200| = 200$, so the company is $\$200$ from break-even. Absolute value answers "how far from zero?"

11 $6 + (-11) + 3 = -5 + 3 = -2$. The team had a net loss of 2 yards.

12   Count the negative factors: there are 4 (even number), so the product is positive. $1 \times 2 \times 5 \times 3 = 30$.

13   Exponent first: $3^2 = 9$. Then multiply: $2 \times 9 = 18$.

14   "r squared" is r^2. "Minus 10" gives $r^2 - 10$.

15   Terms are separated by $+$ or $-$ signs: $6a$, $4b$, and 11.

16   Area $=$ length $\times$ width $= l \times 5 = 5l$.

17   Subtract 9 from both sides: $x = 15 - 9 = 6$.

18 $t \leq 30$ means 30 or less, which is "no more than 30 degrees."

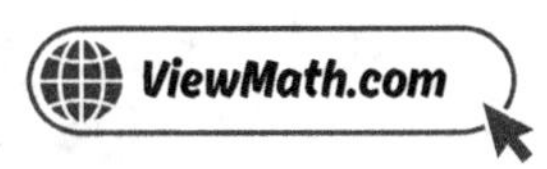

19 The graph represents $x \leq -1$. The value $0 > -1$, so 0 is not a solution.

20 $A = \frac{1}{2} \times 4.5 \times 6 = 13.5 \ m^2$.

21 $V = 3 \times 2 \times 2 = 12 \ ft^3$.

22 For a horizontal side (same y), subtract the x-coordinates and take the absolute value.

23 Length $= |6 - (-4)| = 10$. Width $= |2 - (-3)| = 5$. Area $= 10 \times 5 = 50$ square units.

24 Net A is a valid cross-shaped cube net. Net B has two squares on the same side, which causes overlap. Net C forms a 2×3 block, which doesn't fold into a cube.

25 Even number of values. Middle values are 24 and 28. Median $= (24 + 28) \div 2 = 26$.

26 MAD (Mean Absolute Deviation) is calculated by finding the average of the distances of each data value from the mean.

27 Total $= 2 + 5 + 10 + 8 + 5 = 30$. The 70–79 bar is tallest (10). Students ≥ 70: $10 + 8 + 5 = 23$. Fraction $= 23/30$.

28 Q_1 is the median of the lower half of the data. About 25% of the data falls at or below Q_1.

29 One extreme outlier pulls the mean away from where most of the data lies. The median would be a better choice.

30 There are 9 values, so the median is the 5th value. In order: $20, 24, 31, 35, 3?, 42, 46, 48, 53$. The 5th value must equal 37, so the missing leaf is 7.

Find more at
ViewMath.com/TX-Grade6

✅ Practice Test 8 — Answer Key

1 B

2 Part A: Value $30/line, Plus $25/line. Part B: Value + extra line = $155; Plus −1 line = $130. Part C: Family Plus is cheape

3 15   **4** B   **5** Yes. 3 tons = 6,000 pounds, which is more than 5,500.   **6** C   **7** C

8 D   **9** B   **10** $|-600| = 600$ dollars from zero; a deposit of $600 brings it to $0.   **11** A

12 D   **13** $(2+3) \times 4 = 20$ **14** $3n-8$ **15** 5 and $(y+8)$ **16** $20r+15$ **17** $w=72$

18 C   **19** No. Both shade to the left, but $x < 8$ has an open circle at 8 and $x \leq 8$ has a closed circle at 8.

20 A   **21** $1\ ft^3$ **22** B   **23** C   **24** $348\ in^2$

25 Mean ≈ 13.3, Median $= 6$. The median is better.   **26** C   **27** Mode = 5000, Range = 4000   **28** C

29 Not always.   **30** B

💡 Time to Learn! 💡

Review the explanations below, **especially for the questions you missed**.

Understanding why each answer is correct builds stronger problem-solving skills.

Tip: Circle any questions you got wrong, then read their explanation carefully.

📖 Practice Test 8 — Detailed Explanations

1 Dogs have 4 parts. Each part = 2 animals. Dogs = $4 \times 2 = 8$.

Find more at
ViewMath.com/TX-Grade6

2. A: $\$120 \div 4 = \$30$; $\$150 \div 6 = \$25$. B: $\$120 + \$35 = \$155$; $\$150 - \$20 = \$130$. C: $\$130 < \$155$.

3. $2 \times 5 = 10$ *strawberries, so* $3 \times 5 = 15$ *cups of yogurt.*

4. $35\% = \dfrac{35}{100}$. *GCF of 35 and 100 is 5:* $\dfrac{35 \div 5}{100 \div 5} = \dfrac{7}{20}$.

5. $3 \times 2,000 = 6,000$ *lb.* $6,000 > 5,500$, *so yes.*

6. *The table shows that only the credit card does not use your own money and can charge interest, meaning it involves borrowing.*

7. $k = \frac{35}{5} = 7$. *When* $x = 8$: $y = 7 \times 8 = 56$.

8. *Plan A:* $\$60 \times 4 = \$240$. *Plan B:* $\$75 \times 4 = \$300$. *Difference* $= \$300 - \$240 = \$60$.

9. *The temperatures below zero are* $-3°F$ *and* $-6°F$. *That is 2 temperatures. Zero itself is not below zero.*

10. $|-600| = 600$, *so the balance is $600 from zero. Since* $-600 + 600 = 0$, *a deposit of $600 (the opposite) brings the account to break-even.*

11. $8 - 13 = 8 + (-13) = -5$. *The temperature dropped from 8°C to* $-5°C$.

12. *Count the negative factors: there are 3 (odd number), so the product is negative.* $2 \times 3 \times 4 = 24$, *so the answer is* -24.

13. *Without parentheses:* $2 + 12 = 14$. *With parentheses:* $(2 + 3) \times 4 = 5 \times 4 = 20$.

14. *Start with* n. *Multiply by 3:* $3n$. *Subtract 8:* $3n - 8$.

Find more at
ViewMath.com/TX-Grade6

15. $5(y + 8) = 5 \times (y + 8)$. The factors are 5 and $(y + 8)$.

16. Regular seats: $20r$. VIP seats: 15. Total: $20r + 15$.

17. Multiply by 8: $w = 9 \times 8 = 72$.

18. You can drive at 65 or below, so ≤ 65.

19. The direction is the same, but the circle type differs: open vs. closed.

20. Triangle A: $\frac{1}{2} \times 4 \times 4 = 8$. Triangle B: $\frac{1}{2} \times 5 \times 3 = 7.5$. Triangle A has the greater area.

21. $V = \frac{1}{3} \times \frac{1}{2} \times 6 = \frac{6}{6} = 1$ ft^3.

22. Same y-coordinate, so it is a horizontal distance. $|6 - (-4)| = |10| = 10$ units.

23. Length $= |5 - (-3)| = 8$. Width $= |2 - (-4)| = 6$. Area $= 8 \times 6 = 48$ square units.

24. $SA = 2(15)(6) + 2(15)(4) + 2(6)(4) = 180 + 120 + 48 = 348$ in^2.

25. Mean: $(3 + 4 + 5 + 6 + 7 + 8 + 60) \div 7 = 93 \div 7 \approx 13.3$. Median: 6. The outlier 60 inflates the mean. The median (6) is more typical.

26. The range uses only the maximum and minimum values, so one extreme outlier can make the range very large. The IQR ignores the most extreme values.

27. 5000 has frequency 6 (highest). Range $= 7000 - 3000 = 4000$.

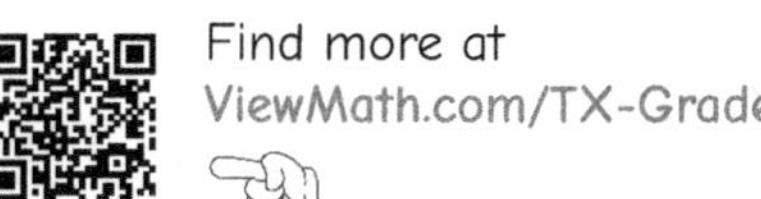

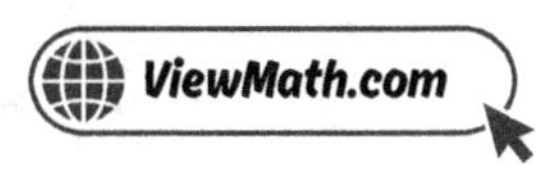

28. Median $= 11$ (6th value). Upper half: $14, 15, 18, 20, 22$. Median of upper half $= 18$. So $Q_3 = 18$.

29. A single outlier can make the range very large while the rest of the data is tightly clustered. The IQR gives a better picture of overall spread. For example, $\{1, 50, 51, 52, 53\}$ has range 52 but most values are close together.

30. Values greater than 50: $51, 53, 56, 62$. That is 4 values.

✅ Practice Test 9 — Answer Key

1. C 2. B 3. B 4. A 5. B 6. C 7. B 8. Part A: \$75; Part B: \$5

9. B

10. Part A: $|-30| = 30$, $|45| = 45$, $|-60| = 60$, $|15| = 15$, $|-45| = 45$; Part B: Book Barn (60 dollars); Part C: F

11. A 12. B 13. B 14. B 15. C 16. C 17. C 18. B 19. B 20. C

21. B 22. C 23. A 24. C 25. B 26. C 27. B 28. B 29. B 30. 28

💡 Time to Learn! 💡

Review the explanations below, **especially for the questions you missed**.

Understanding why each answer is correct builds stronger problem-solving skills.

Tip: Circle any questions you got wrong, then read their explanation carefully.

📖 Practice Test 9 — Detailed Explanations

Find more at
ViewMath.com/TX-Grade6

1. Total parts $= 1 + 3 + 2 = 6$. Each part $= 18 \div 6 = 3$. Blue $= 3 \times 3 = 9$.

2. $\$5.40 \div 12 = \$0.45$ per juice box.

3. $3 \times 2 = 6$, so $5 \times 2 = 10$.

4. Tax: $0.06 \times \$50 = \$3$. Total: $\$50 + \$3 = \$53$.

5. $3 \times 8 = 24$ pints.

6. A streaming subscription is entertainment, which is a want. Clothing, water, and medical care are needs.

7. Table 1: $\frac{3}{1} = 3$, $\frac{6}{2} = 3$, $\frac{12}{4} = 3$ (constant). Table 2: $\frac{3}{1} = 3$, $\frac{7}{2} = 3.5$ (not constant). Only Table 1 is proportional.

8. Part A: Friend A $= \$40 \times 5 = \$200$. Friend B $= \$25 \times 5 = \$125$. Difference $= \$200 - \$125 = \$75$. Part B: $I = 125 \times 0.04 \times 1 = \$5$.

9. Underground means below ground level, which is negative. 120 feet underground is -120.

10. Part A: $|-30| = 30$, $|45| = 45$, $|-60| = 60$, $|15| = 15$, $|-45| = 45$. Part B: Book Barn has the greatest absolute value (60), so it is farthest from break-even. Part C: Pet Palace ($\$45$) and Green Garden ($-\$45$) both have an absolute value of 45. They are the same distance from zero because 45 and -45 are opposites.

11. First, $(-3) + (-5) = -8$ (same signs, add: $3 + 5 = 8$, keep negative). Then $-8 + 2 = -6$ (different signs: $8 - 2 = 6$, keep negative).

12. Both factors are positive. Positive $\times$ positive $=$ positive. $6 \times 8 = 48$.

13. Parentheses: $4 + 4 = 8$. Divide: $48 \div 8 = 6$.

Find more at
ViewMath.com/TX-Grade6

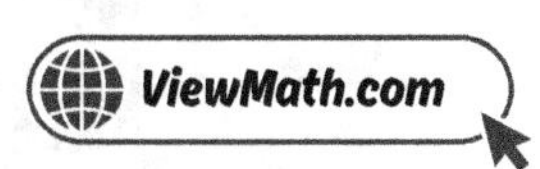

14 *"7 less than y" means subtract 7 from y: $y - 7$. The order is reversed from how it reads.*

15 *The expression $2(a + 9)$ has two factors: 2 and $(a + 9)$. The variable a by itself is not a factor of the whole expression — it is part of the group $(a + 9)$.*

16 *Hourly earnings: $8h$. Plus tip: $8h + 15$.*

17 *Divide both sides by 8: $w = 72 \div 8 = 9$.*

18 *"Below 0" means less than 0: $t < 0$. (Not equal to 0, since it says "below.")*

19 *"More than 75" is $t > 75$: open circle (not including 75), shade right (greater values).*

20 $A = \frac{1}{2} \times \frac{1}{2} \times \frac{1}{4} = \frac{1}{16}$ *ft^2.*

21 *Box A volume $= 6 \times 4 \times 5 = 120$. Box B: $120 = 10 \times 3 \times h = 30h$, so $h = 4$.*

22 *Same x-coordinate, so it is a vertical distance. $|5 - (-3)| = |8| = 8$ units.*

23 *Base $= 10$. Height $= 4$. Area $= \frac{1}{2} \times 10 \times 4 = 20$ square units.*

24 *A cube has 6 equal faces. $SA = 6 \times 5^2 = 6 \times 25 = 150$ in^2.*

25 *There are 15 data points. The median is the 8th value. Counting from 0: 2 zeros, 3 ones (5 so far), then 4 twos (9 so far). The 8th value is 2.*

26 *Adding 5 to every value shifts the max and min by the same amount, so the difference (range) stays unchanged.*

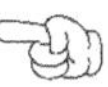

Find more at
ViewMath.com/TX-Grade6

27   *The 25–49 bar has height 8, the tallest bar. It is the modal interval.*

28   *Team X median = 70, Team Y median = 75. Team Y is higher.*

29   *X: $Q_1 = 20$, $Q_3 = 60$, IQR = 40. Y: $Q_1 = 37$, $Q_3 = 43$, IQR = 6. Y is much smaller.*

30   *The values are $20, 24, 28, 32, 36$. Sum $= 20 + 24 + 28 + 32 + 36 = 140$. Mean $= 140 \div 5 = 28$.*

☑ Practice Test 10 — Answer Key

1   C   **2**   D   **3**   B   **4**   B   **5**   B

6   Part A: 15%. Part B: Savings $200, Clothing $100, Electronics $125, Charity $75. Part C: Savings = 45% of $500 = $225; E

7   Yes; $y = 8x$ **8**   6 weeks   **9**   B   **10**   C

11   Part A: Arrow 1 $= -4$, Arrow 2 $= +7$, Arrow 3 $= -5$.   Part B: $(-4) + 7 + (-5)$.   Part C: -2.   **12**   D

13 $5^2 = 25$ square units   **14**   C   **15**   D   **16** $3 + 2m$ **17**   A   **18** $p \leq 12$ and $w \leq 2{,}000$

19   C   **20**   B   **21**   B   **22**   C   **23**   24 square units   **24**   C   **25**   B   **26**   Data set B

27   B   **28**   B   **29**   C   **30**   B

Find more at
ViewMath.com/TX-Grade6

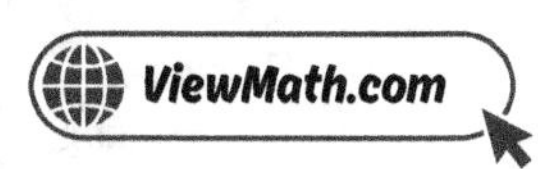

💡 Time to Learn! 💡

*Review the explanations below, **especially for the questions you missed.***

Understanding why each answer is correct builds stronger problem-solving skills.

Tip: *Circle any questions you got wrong, then read their explanation carefully.*

📖 Practice Test 10 — Detailed Explanations

1 Apple has 3 parts, orange has 5 parts. Total parts = 8. Total ounces = $8 \times 4 = 32$.

2 Clean & Fresh: $\$9.60 \div 32 = \$0.30$/load. Sparkle Wash: $\$12.00 \div 48 = \$0.25$/load. Sparkle Wash is cheaper.

3 $2 : 3$ multiplied by 3 gives $6 : 9$. The other pairs do not simplify to $2 : 3$.

4 Sale price: $\$45 - 0.20 \times \$45 = \$45 - \$9 = \$36$. Tax: $0.08 \times \$36 = \$2.88$. Total: $\$36 + \$2.88 = \$38.88$.

5 Converting 5 feet to inches uses the conversion ratio $\frac{12\ in}{1\ ft}$.

6 A: $100\% - 40\% - 20\% - 25\% = 15\%$. B: $0.40 \times 500 = \$200$; $0.20 \times 500 = \$100$; $0.25 \times 500 = \$125$; $0.15 \times 500 = \$75$. C: New Savings = 45%: $0.45 \times 500 = \$225$. New Electronics = 20%: $0.20 \times 500 = \$100$.

7 There is no extra fee, so the total cost $y = 8x$. This is in the form $y = kx$, making it proportional with $k = 8$.

8 Weekly savings = $0.50 \times \$40 = \$20$. Number of weeks = $\$120 \div \$20 = 6$ weeks.

9 On a horizontal number line, negative numbers are always to the left of zero and positive numbers are to the right.

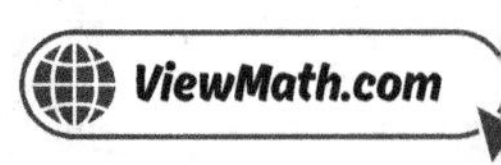

10 A number and its opposite are the same distance from zero on different sides, so they cancel each other out. For example, $6 + (-6) = 0$ and $-9 + 9 = 0$.

11 Arrow 1 goes from 0 to -4: that is -4. Arrow 2 goes from -4 to 3: that is $+7$. Arrow 3 goes from 3 to -2: that is -5. The expression is $(-4) + 7 + (-5) = 3 + (-5) = -2$.

12 Descending 8 feet is -8 per minute. Over 7 minutes: $7 \times (-8) = -56$ feet.

13 The square has side length 5, so area $= 5^2 = 25$ square units.

14 Half of n is $\frac{n}{2}$. "6 more" means add 6: $\frac{n}{2} + 6$.

15 $x = 1 \cdot x$, so the coefficient is 1. The expression has 2 terms: x and 5.

16 Flat fee: \$3. Per-mile cost: $2m$. Total: $3 + 2m$.

17 Subtract 3.5: $x = 10 - 3.5 = 6.5$.

18 "Maximum 12 persons" means $p \leq 12$. "Maximum weight 2,000 lbs" means $w \leq 2,000$.

19 Open circle at 0 means 0 is not included. Shading right means greater than: $x > 0$.

20 $24 = \frac{1}{2} \times 8 \times h$, so $24 = 4h$, which gives $h = 6$ ft.

21 $V = 5 \times 3 \times 4 = 60$ cm^3.

22 The base goes from $(0,0)$ to $(8,0)$. Distance $= |8 - 0| = 8$ units.

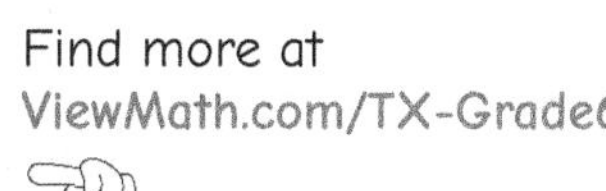
Find more at
ViewMath.com/TX-Grade6

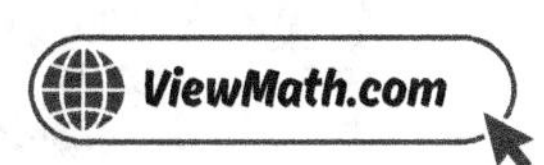

23 Base $= |5 - (-3)| = 8$. Height $= |4 - (-2)| = 6$. Area $= \frac{1}{2} \times 8 \times 6 = 24$ square units.

24 The shaded face is $8\ cm \times 5\ cm = 40\ cm^2$.

25 Original mean: $(58 + 60 + 62 + 64) \div 4 = 61$. New mean: $(58 + 60 + 62 + 64 + 82) \div 5 = 326 \div 5 = 65.2$.

26 A larger MAD means values are farther from the mean on average. Data set B's MAD of 10 indicates more variability than A's MAD of 2.

27 7 appears 2 times, 8 appears 4 times, 9 appears 3 times, 10 appears 1 time. The mode is 8.

28 From the box plot: $Q_1 = 15$, $Q_3 = 30$. $IQR = 30 - 15 = 15$.

29 For symmetric data, the mean is a good center. For skewed data, the median better represents the typical value since it is not pulled by the tail.

30 Stem 5 has 5 leaves, which is more than any other stem.

Well done checking your answers!

Keep practicing to strengthen your skills.

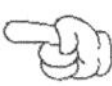

Find more at
ViewMath.com/TX-Grade6

Author's Final Note

I hope you enjoyed this book as much as I enjoyed writing it. Whether you are a student working through the material, a parent supporting your child's learning, or a teacher guiding your class, I have tried to make this book as clear and engaging as possible. I hope I have succeeded. If you have any suggestions for improvement, please let me know. I would love to hear from you.

The accuracy of calculations is very important to me. We have done our best, but I also expect that I have made some minor errors. Constant improvement is the name of the game. If you find any errors, please let me know. I will fix them in the next edition.

For students: Your learning journey does not end here. I have written a series of books to help you learn math. Make sure you browse through them. I especially recommend workbooks and practice tests to help you prepare for your exams.

For parents: Thank you for investing in your child's education. I encourage you to explore the companion resources available online to help support your child outside the classroom.

For teachers: Thank you for the invaluable work you do every day. I hope this book serves as a useful resource in your classroom. Feel free to reach out if you have suggestions or would like to discuss how best to use this book with your students.

I also enjoy reading your reviews. If you have a moment, please leave a review on where you found this book. It will help others find this book. If you have any questions or comments, please feel free to contact me at DrNazari@ViewMath.com.

And one last thing: Remember to use online resources for additional help. I recommend using the resources on https://ViewMath.com You can find video lessons, practice problems, and more. You can also use the online companion for this book to track your progress and access additional resources.

Wishing all students the best in their studies, parents every success in supporting their children, and teachers continued inspiration in their classrooms!

Dr. A. Nazari

Great Job! Keep Learning with ViewMath!

*Keep up the great work! Visit **viewmath.com/TX-Grade6** for free lessons, quizzes, and more.*

Study Guide

Workbook

Step-by-Step

3 Practice Tests

5 Practice Tests

7 Practice Tests